Tidal Current Charts of Southeast Alaska

Grenville Channel to Skagway

Second Edition

By
Randel Washburne

Edited by
David Burch

ACKNOWLEDGMENTS

We remain grateful to Robert Hale, publisher of the first edition, for his contribution to the concept and goals of this atlas, and to Joan Kohl for her design of that book and the charts.

New supporting graphics and tables are by Tobias Burch. Front cover photo, "Looking back to Little Port Walter," by NOAA Commander John Bortniak, is from the NOAA public Photo Library.

Paintings by Randel Washburne, based on the black and white photographs of the original 1977 edition of Joe Upton's *Alaska Blues* (Epicenter Books, Kenmore, WA, 2008)—a book that reveals the mystical qualities of the Inside Passage that have drawn the author and others into this remarkable waterway.

We also wish to acknowledge the influence of the 2016 Race to Alaska on motivating this second edition of the Atlas. Although some vessels now have electronic chart displays that illustrate current flow very nicely over a broad region, there are still many vessels that do not, or cannot, take advantage of those resources. Our work with the winning Team MAD Dog Racing was just one example of an open boat making long passages with essentially none of the standard navigation equipment. Many other vessels in these waters who could use electronic aids, prefer the simplicity and dependability of just looking at the traditional current tables and turning to a printed picture of the current flow across the waterway.

978-0-914025-54-2

Published by

Starpath Publications

3050 NW 63rd Street, Seattle, WA 98107

Manufactured in the United States of America

www.starpathpublications.com

INTRODUCTION

Background

These tidal current charts of this atlas are presented in the style of the former NOAA Graphical Current Charts, which were available up until 1990. Those charts covered nine waterways around the contiguous US, including Puget Sound, San Francisco Bay, and Chesapeake Bay. All nine books were discontinued after forty years of continuous publication because better data and other resources had become available. Navigators who grew up using them were sorry to see them go.

These southeast Alaska current charts return to that traditional format. They show chartlets of the region, with a current arrow plotted at each location where daily NOAA tidal current data are available. The area is divided into three sections (North, Central, and South) and there are thirty-six charts for each section representing different stages of a tidal cycle. The size and direction of the arrows on the individual charts represent the average flow expected for the valid time of that chart. Instructions are provided for selecting the proper chart for specific times and dates. The choice is based on the daily values of the tidal currents at Wrangell Narrows, which must be obtained separately as explained in the Instructions.

A few channels in this waterway are restricted enough to create an imbalance in the tide height at each end when the tide turns. This creates an extra force on the current flow, called a hydraulic head, which in turn causes them to behave somewhat differently than other channels. These few have been separated out into a section called SELECTED CHANNELS, and they are scaled to the Sergius Narrows data, rather than Wrangell Narrows. These special channels are marked in shaded areas on the main chart pages.

Current Flow Overview

Tidal currents in southeast Alaska waters can reach 10 knots in certain channels, with eddies, overfalls, and tide rips that could be dangerous to any vessel. The currents in most passages, however, are far slower and less hazardous. Nevertheless, theses currents can greatly affect speed made good and fuel consumption. It is not unusual to find a channel more than 20 miles long with tidal current speeds exceeding three knots for several hours.

This current atlas provides an easy way to plan voyages to take advantage of favorable currents and avoid unfavorable currents. Arrows show the direction and speed of current to be expected during every hour. Current atlas predictions have been checked extensively against the actual NOAA predictions, and have been found remarkably accurate. Even so, the prudent mariner will not rely on the atlas predictions alone. The annual NOAA tidal current tables provide more precise information, and they provide the best information about the timing of slack waters. Local conditions, such as higher or lower than normal atmospheric pressure, sustained winds, and river runoff will affect currents. Unusually strong currents caused by these conditions will be understated in this atlas. Thus, this atlas should be used for typical conditions, and where extreme precision is not required. Used in this way, this atlas will be a useful navigation tool.

Data Sources

Timing and speeds of currents were developed from data in the *Tidal Current Tables: Pacific Coast of North America and Asia* ("NOAA tidal current tables") for 1989; *U.S. Coast Pilot, Pacific Coast, Alaska: Dixon Entrance to Cape Spencer*; and *Sailing Directions, British Columbia Coast (North Portion)*. There is no evidence as of 2016 that the annual average currents in this waterway have varied since the original compilation of this data.

Most of the arrows on the current atlas maps are positioned at secondary current stations listed in Table 2 of the NOAA tidal current tables. The times of slack and of maximum tidal current flow, and the directions of flow, were determined from this NOAA information, using Wrangell Narrows and Sergius Narrows stations. As a simplifying procedure, secondary stations based on North Inian Pass predictions were adjusted to apply to Wrangell Narrows. This adjustment introduces error of not more than one hour in a few cases.

In a number of channels, NOAA provides more secondary stations than can be displayed on the current atlas maps scale. In these situations, stations were chosen that appeared to best represent the local situation and which seemed most consistent with other stations in the channel. It is likely that flows suggesting eddies or short-distance acceleration around land features were omitted in these confined waters. At some locations, no data were available from either the NOAA tidal current tables or, in British Columbia, from the Canadian Hydrographic Service current tables. In these cases, information about current behavior was drawn from the U.S. Coast Pilot or from the Canadian Sailing Directions. These descriptions are much less precise about timing and speeds, so the flows are shown in the atlas by dashed "Speed Unknown" arrows. The user should understand that "Speed Unknown" arrows are estimates of the situation only and their locations do not necessarily coincide with known tabulated current stations.

Averaging and Rounding

To develop this current atlas, an entire year's tidal current speeds were plotted. The range of current speeds was divided into thirds, representing small, medium, and large current flows. The current speeds for each third were averaged, and these averages became the basis for the current speeds shown on the charts. Thus, the current atlas overestimates current speeds at the bottom of each third, and underestimates at the top.

Times were rounded to the nearest whole hour. If, for example, a particular secondary station in the NOAA tidal current tables is predicted to turn to flood 35 minutes before Wrangell Narrows, the atlas shows that station turning one hour earlier than Wrangell Narrows. A station turning 25 minutes before Wrangell Narrows is shown as turning the same hour as Wrangell Narrows.

The navigator should note that each chart shows the general conditions to be expected during the hour that begins at the time in question. If, for example, it were determined that chart 23 is appropriate at 0600 on a particular day, that chart would show the average currents found between 0600 and 0700. It would show most closely the currents to be expected at 0630. For these reasons, in addition to an understanding of weather, local conditions, and any local knowledge that may apply, the current atlas should be used to show general trends and the progression of tidal current flows throughout a tidal cycle. Refer to the NOAA tidal current tables for the highest precision at specific locations.

Daily Tidal Current Data

To use this atlas, you will need up-to-date current data for Wrangell Narrows or Sergius Narrows, as appropriate. The primary printed source are the NOAA tidal current tables. The printed version is available at marine outlets or from several retailers. That book, however, covers all stations in that broad area, whereas only two stations are needed for this atlas.

Alternatively, the NOAA annual data for Wrangell Narrows and Sergius Narrows can be downloaded as a free pdf from

www.tidesandcurrents.noaa.gov,

using station ID "SEA0103 Depth 4 feet" for Wrangell Narrows.

There is a link there that creates the annual pdf of the same data that appears in the official printed publication. Most locations in the atlas are based on Wrangell Narrows. Only a few selected channels use Sergius Narrows (station ID "SEA0202 Depth 18 feet"). They are presented in the SELECTED CHANNELS section.

The NOAA printed edition will always be in standard time (LST) so you have to add 1 hour to get Alaska Daylight Time (AKDT) when in effect. But when you create your own pdf table from the link above, you can choose to have the times automatically corrected to AKDT when in effect, marked LST/LDT. You can also choose 12h or 24h times, and the latter will usually be more convenient for this application. Though rarely a practical factor, the downloaded tables are adjusted to best known data at the time of download, which could be an improvement over the printed annual edition.

Table data are listed with slack water times in the left column and the times of maximum current flow in the middle column, followed on the right with the value of the maximum current. The ebb currents are traditionally labeled and graphed as a minus value, but this has no significance. A sample is shown in Figure 1.

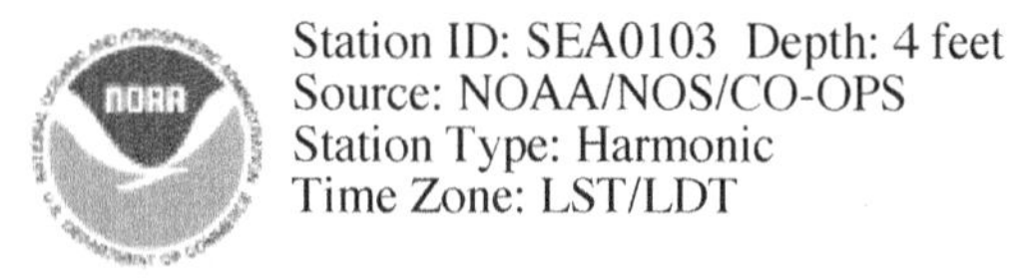

Station ID: SEA0103 Depth: 4 feet
Source: NOAA/NOS/CO-OPS
Station Type: Harmonic
Time Zone: LST/LDT

NOAA Tidal Current Predictions

Wrangell Narrows (off Petersburg), 2016

Latitude: 56.8150° N Longitude: 132.9628° W
Mean Flood Dir. 246° (T) Mean Ebb Dir. 62° (T)
Times and speeds of maximum and minimum current, in knots

July

	Slack	Maximum			Slack	Maximum	
	h m	h m	knots		h m	h m	knots
1 F	 05:42 11:48 18:00 23:48	02:30 08:54 15:00 21:06	-2.5E 3.5F -2.0E 3.5F	**16** Sa	 06:06 12:06 18:12	02:54 09:12 15:24 21:18	-1.8E 2.8F -1.3E 2.7F
2 Sa	 06:36 12:48 18:54	03:24 09:42 15:54 22:00	-2.9E 3.9F -2.3E 3.7F	**17** Su	00:00 06:48 12:54 19:00	03:42 10:00 16:06 22:06	-2.1E 3.1F -1.6E 2.9F

August

	Slack	Maximum			Slack	Maximum	
	h m	h m	knots		h m	h m	knots
1 M	00:30 07:12 13:36 19:30	04:00 10:18 16:30 22:36	-2.9E 4.0F -2.4E 3.6F	**16** Tu	00:18 07:06 13:12 19:24	03:54 10:18 16:24 22:24	-2.3E 3.4F -1.9E 3.2F
2 Tu ●	01:18 08:00 14:18 20:18	04:48 11:06 17:18 23:18	-3.2E 4.2F -2.6E 3.8F	**17** W	01:06 07:42 13:54 20:00	04:36 10:54 17:00 23:06	-2.6E 3.8F -2.3E 3.5F

September

	Slack	Maximum			Slack	Maximum	
	h m	h m	knots		h m	h m	knots
1 Th ●	01:54 08:24 14:30 20:42	05:12 11:30 17:36 23:42	-3.0E 4.2F -2.7E 3.9F	**16** F ○	01:30 07:54 14:00 20:18	04:54 11:06 17:12 23:24	-3.0E 4.3F -3.0E 4.2F
2 F	02:30 08:54 15:00 21:12	0[illegible]48 12:0[illegible] 18:12	-3.0E 4.1F -2.7E	**17** Sa	02:18 08:30 14:42 20:54	05:30 11:42 17:48	-3.3E 4.5F -3.3E

	Slack	Maximum	
	h m	h m	knots
1 Th ●	01:54 08:24 14:30 20:42	05:12 11:30 17:36 23:42	-3.0E 4.2F -2.7E 3.9F

Figure 1. *A sample from the free pdf download. Using Sept 1, 2016 as a sample date, we see that the sequence of current times in AKDT are:*

0154	*Slack*
0512	*3.0 kts Ebb*
0824	*Slack*
1130	*4.2 kts Flood*
1430	*Slack*
1736	*2.7 kts Ebb*
2042	*Slack*
2342	*3.9 kts Flood.*

INSTRUCTIONS

Follow this procedure to determine which chart to use for a specific time and date.

Step 1. Find the Wrangell Narrows current table entries for the day in question. In the left-hand column, find the time for slack water preceding or coinciding with the hour of interest. Round this slack time to the nearest hour.

Step 2. Find the maximum speed of the current following the slack determined in Step 1. This speed will be on the same line, at the right edge of the day's table—or on the first line of the next day's table if the slack occurs at the end of the day. Note whether the current is a Flood (F) or an Ebb (E).

Step 3. Using the maximum current found in Step 2, determine which row of Table 1 to use for finding the chart number. For example, a Wrangell Narrows maximum Ebb of 3.4 knots would use the row marked "3 – 4" in the Ebb column.

Step 4. Compute the number of hours that elapsed since the slack found in Step 1. This will determine the column to use to find the correct chart.

For example, if the time of interest is 1500 and the most recent slack was at 1300, the elapsed time is two hours. Read across the row found in Step 3 to find the chart that is recommended for the number of hours since most recent slack.

In other words, if the time of interest is 1500, and this time is two hours later than the most recent slack at Wrangell Narrows, which is building to a maximum of 3.4 kts Ebb, then Table 1 tells us that 1500 corresponds to chart 27. Then from Table 1 we know immediately that 1600 would be chart 28, and 1700 would be chart 29.

Example 1. Find the proper chart for 1000 AKDT on Sept 1, 2016.

Step 1. Slack preceding 1000 is at 0824, which rounds to 0800.

Step 2. Max current following 1000 is 4.2 kts Flood.

Step 3. The 4.2 kts Flood tells us to use the top row in the Flood column, marked "> 4."

Step 4. Our time of 1000 is 2h after the most recent slack at 0800, so chart 3 applies to 1000 on Sept 1, 2016, being the best overview at 1030. Then chart 4 would be 1100, chart No. 5, 1200.

Table 1. Chart Number Based on Slack Time and Max Speed

Wrangell Narrows Max. Speed (kts)	*Hours after most recent slack*					
Flood	*0h*	*1h*	*2h*	*3h*	*4h*	*5h*
> 4	1	2	3	4	5	6
3 – 4	7	8	9	10	11	12
< 3	13	14	15	16	17	18
Ebb	*0h*	*1h*	*2h*	*3h*	*4h*	*5h*
> 4	19	20	21	22	23	24
3 – 4	25	26	27	28	29	30
< 3	31	32	33	34	35	36

An alternative way to select the right chart is on the next page.

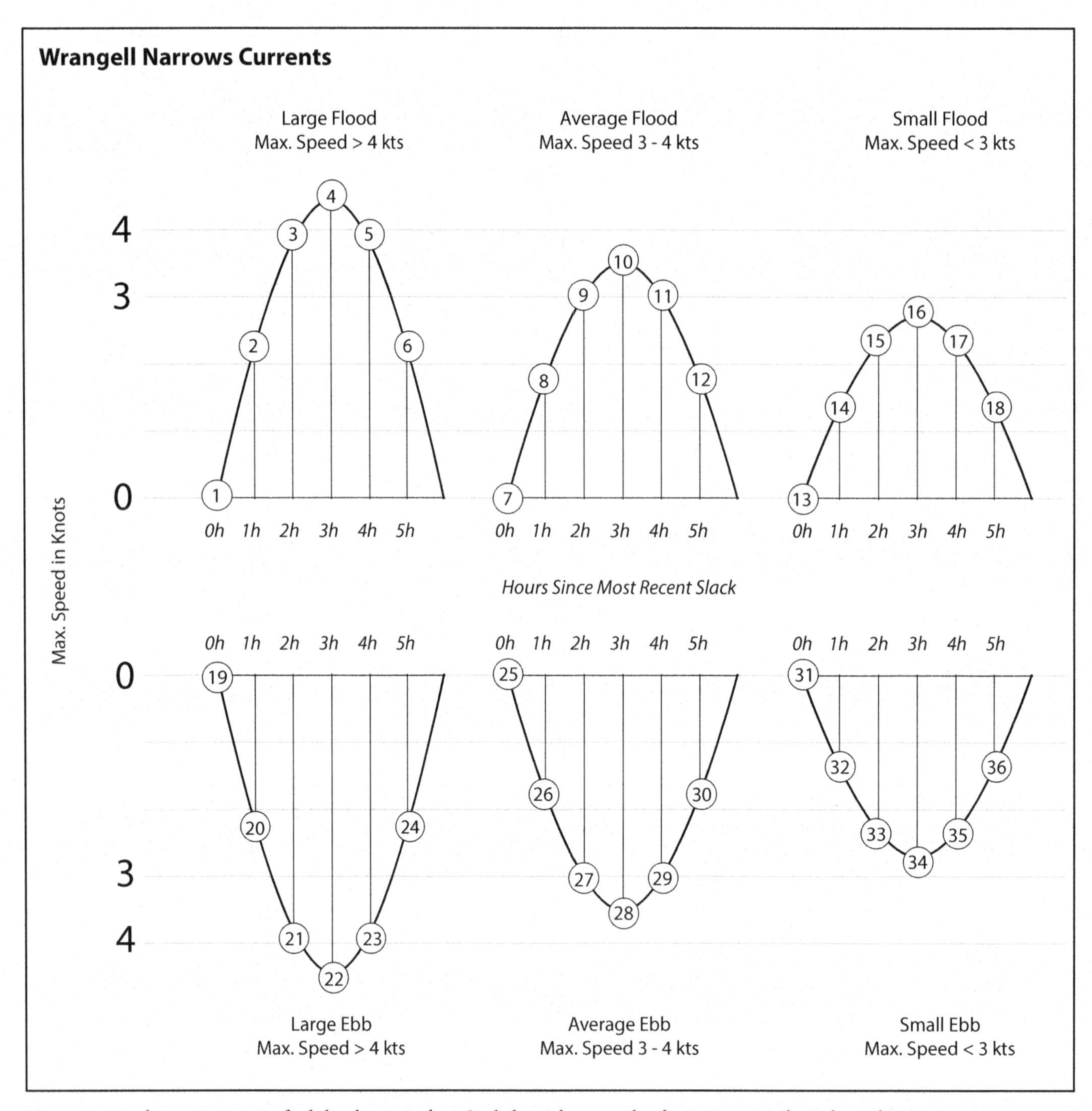
Wrangell Narrows Currents
Large Flood
Max. Speed > 4 kts
Average Flood
Max. Speed 3 - 4 kts
Small Flood
Max. Speed < 3 kts
4
3
0
Max. Speed in Knots
0h 1h 2h 3h 4h 5h
Hours Since Most Recent Slack
1 2 3 4 5 6
7 8 9 10 11 12
13 14 15 16 17 18
19 20 21 22 23 24
25 26 27 28 29 30
31 32 33 34 35 36
Large Ebb
Max. Speed > 4 kts
Average Ebb
Max. Speed 3 - 4 kts
Small Ebb
Max. Speed < 3 kts

Figure 2. *An alternative way to find the chart number. Circled numbers are the charts to use at the indicated times.*

Alternative Way to Find the Proper Chart

The Table 1 numeric instructions can be presented graphically as in Figure 2. This shows that if your time of interest coincides with a slack water at Wrangell Narrows, then the right chart for that time would be either 1, 7, or 13 for a building Flood, or 19, 25, or 31 for a building Ebb. The right choice depends on the expected maximum value of the current. A large ebb (max current > 4 kts) cycle starts on chart 19 and progresses hourly through 24. After chart 24 you will be at slack again, this time preceding a flood. In this example, the next chart after 24 will be 1, 7, or 13. If the coming flood is a small one (< 3 kts), then the chart after 24 would be 13, then proceeding though that flood cycle to chart 18.

Extending this approach to other times, if you care about the currents right now and this is 3h after the slack at Wrangell Narrows within an average flood cycle (maximum current of 3-4 kts), then chart 10 describes present conditions, chart 11 is one hour later, and chart 12 two hours later. At 3h later you would have to see what type of ebb is building to determine the next set.

Video examples of the use of the Atlas are presented at www.starpath.com/videos.

Special Cases

This current atlas is intended to facilitate route planning by presenting graphic views of the overall flow patterns. The indicated speeds should be consistent with the NOAA predictions to within the definitions of the arrow sizes in most cases, but there can be exceptions, both in speed and in times. First, the NOAA predictions themselves have an uncertainty in normal conditions of ±0.3 kts (RMS) in the current speed at any time. Slack and max current times are uncertain by about ± 10 min. When either one of these extremes occurs, it could in principle change the correct chart number. This is unlikely, but possible.

Environmental factors can also affect the results in an unpredictable manner. A strong (>17 kts) wind blowing in the ebb direction for 12 hr or more, for example, will enhance the strength of the ebb and expand its duration, while shortening the flood cycle and weakening its maximum current. This is an effect we can keep in mind when on the water, but it cannot be built into the current predictions.

There is also the influence of the mixed-diurnal tidal cycle, which has two high waters and two low waters each day. The relative heights of these peaks change throughout the month, which can, once a month or so, distort the current flow pattern as shown in Figure 3. Aug 13, 2016, for example, had an exceptionally long ebb cycle, with slack to slack of 7h 06m, instead of the normal average of 6h. These exceptions are easy to spot in the Tidal Current Tables as they usually call for an extra line of data marked by " ***** " in the download version and by "†" in the annual printed edition. The Extra Data are at the back of the Tables (Figure 4).

The relative flow at various locations shown on the charts can still be useful guide in these conditions, but with larger uncertainties, and individual stations should be checked for the best predictions. In this example, for the start of the small (irregular) ebb at 2130 on Aug 12 we would use chart 31, and this ebb cycle ends with the start of a (regular) small flood at 0436 on Aug 13, which would use chart 13, leading into a normal chart selection procedure for later times.

With the two endpoint charts correct, we can improve the selection of the midpoints for the extended ebb by rescaling the chart intervals. Instead of the regular 1h steps, we note that 7h 6m/6 is 428m/6 = 1h 11m per chart. The biggest effect on chart selection occurs at the end of the cycle. In this example, chart 36, just before the next ebb, would apply to 2130 Aug 12 + 5h 55m = 0325 Aug 13, which will show the currents weakening, getting ready to turn.

It helps in these special cases—and indeed all applications of the charts—to recall that the charts best represent the times halfway between the assigned chart times. A chart assigned to 1300, for example, is intended to describe the waterway from 1300 to 1400, with the most accurate representation applying to 1330. If you notice, for example, that one particular channel on a given chart is flooding on one chart and then ebbing on the next, then it turned at about the assigned time of the latter chart, which is halfway between the two optimum times of the successive charts.

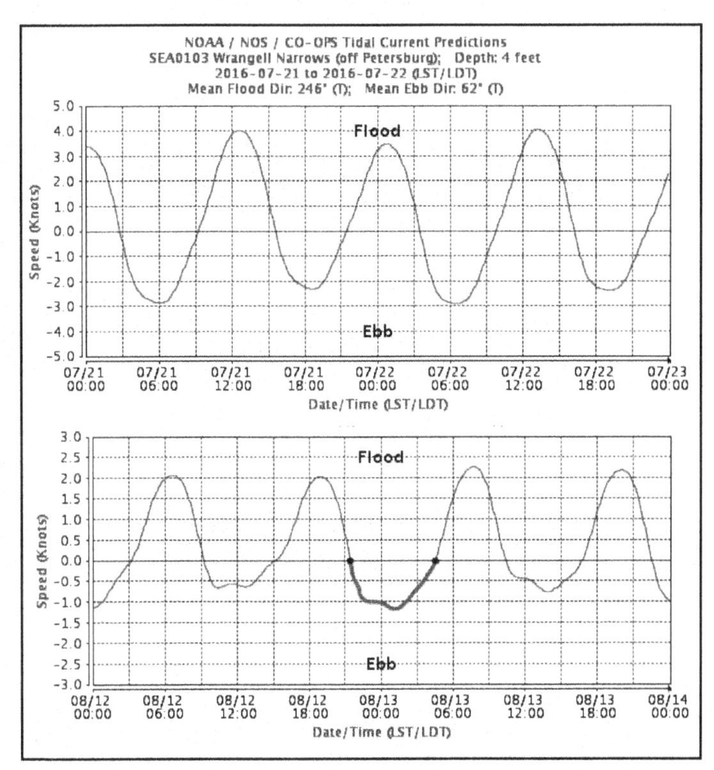
NOAA / NOS / CO-OPS Tidal Current Predictions
SEA0103 Wrangell Narrows (off Petersburg); Depth: 4 feet
2016-07-21 to 2016-07-22 (LST/LDT)
Mean Flood Dir: 246° (T); Mean Ebb Dir: 62° (T)
Speed (Knots)
Flood
Ebb
Date/Time (LST/LDT)
Speed (Knots)
Flood
Ebb
Date/Time (LST/LDT)

Figure 3. *Plot of a normal current cycle (top) compared to an extended ebb cycle (Bottom). From tidesandcurrents.noaa.gov.*

Day	Slack	Maximum	Knots
12 F		00:00	-1.1E
	03:18	06:36	2.1F
	09:18	10:24	-0.7E
		12:36	-0.6E
	15:06	18:54	2.0F
	***************		
13		01:12	-1.2E
	04:36	07:42	2.3F

Extra Data
Friday, August 12, 2016

Slack h m	Maximum h m	knots
21:30	23:06	-1.0E

Figure 4. *Current data for an extended cycle, using Extra Data from the back of the Tidal Current Tables.*

For the even rarer cases when the cycle is closer to 5h than the average of 6h, then the same procedure for scaling the chart times can be applied, and each chart interval will be about 50m per chart rather than 1h. Again, both cases are rare, and the effect would mainly impact the optimum choice of the last chart before the next slack.

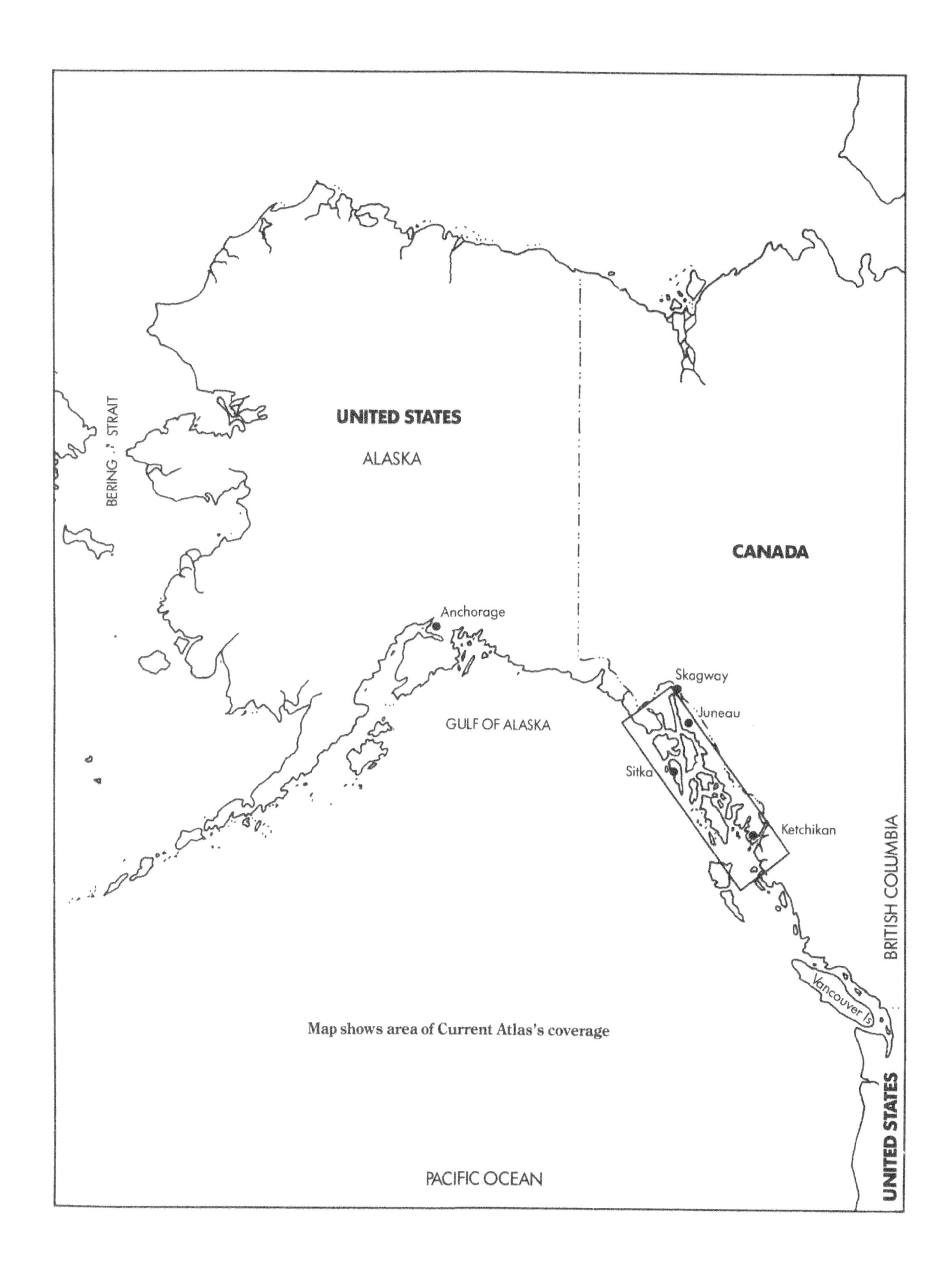
UNITED STATES
ALASKA
BERING STRAIT
CANADA
Anchorage
Skagway
Juneau
Sitka
Ketchikan
GULF OF ALASKA
BRITISH COLUMBIA
Vancouver Is
UNITED STATES
PACIFIC OCEAN

Map shows area of Current Atlas's coverage

Composite of pages at right shows how the current atlas is divided into North, Central, and South sections. Areas of overlap are indicated. North-South orientation was changed as needed to align maps vertically on pages. Locations based on Sergius Narrows predictions are shown by dark squares.

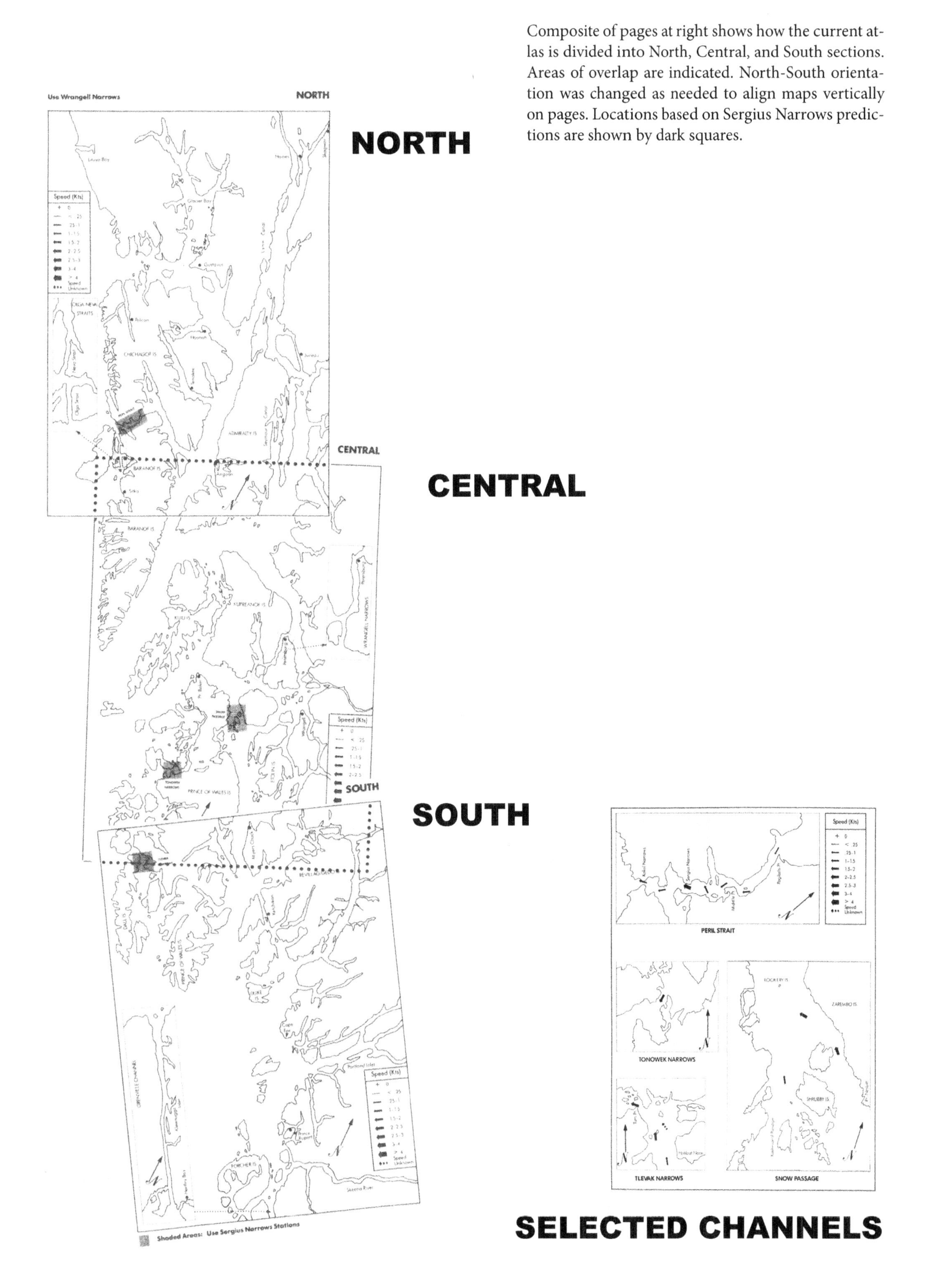

Use Wrangell Narrows
NORTH
NORTH
Speed (Kts)
CENTRAL
CENTRAL
Speed (Kts)
SOUTH
SOUTH
Speed (Kts)
Shaded Areas: Use Sergius Narrows Stations
Speed (Kts)
PERIL STRAIT
TONOWEK NARROWS
TLEVAK NARROWS
SNOW PASSAGE
SELECTED CHANNELS

NORTH

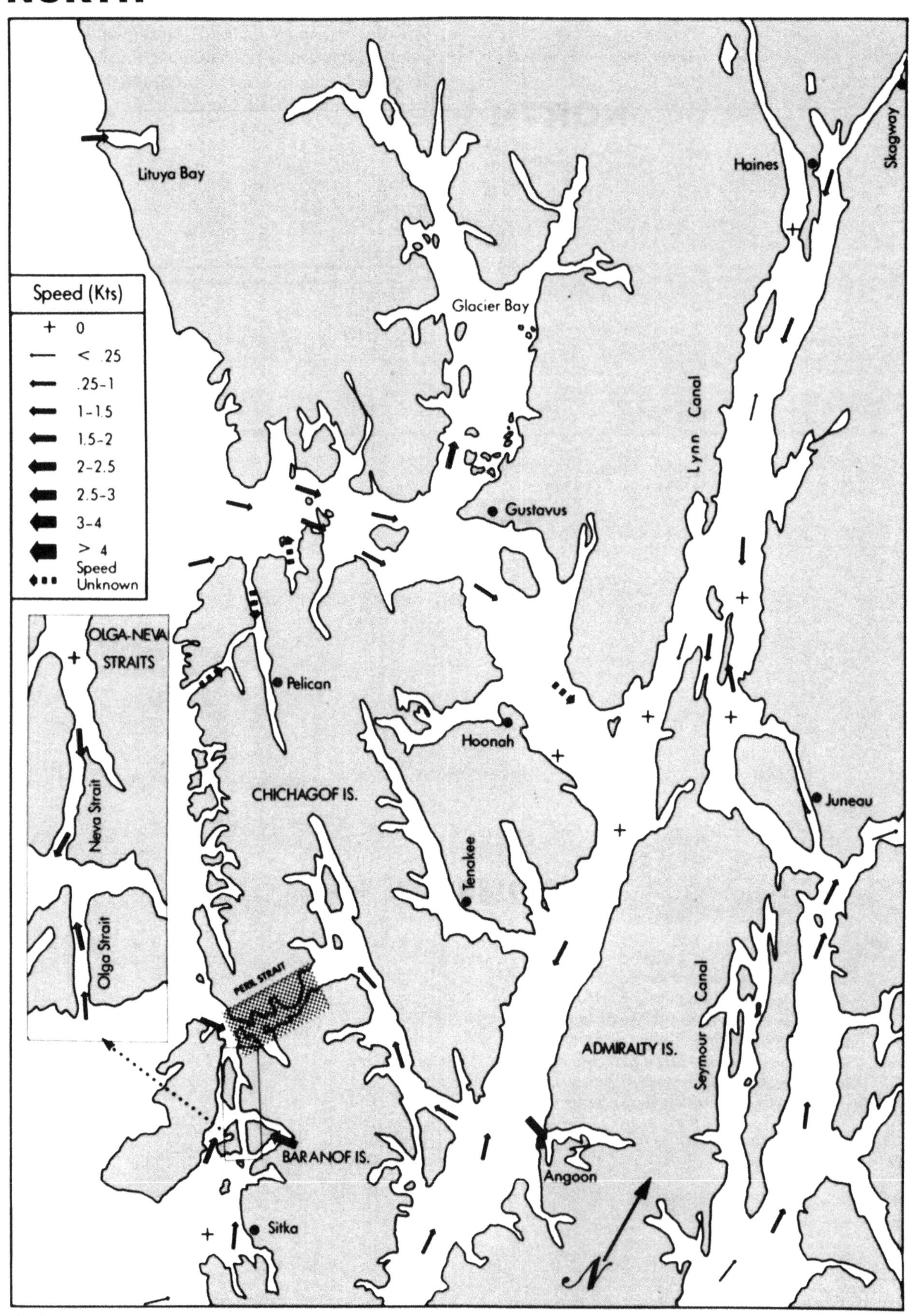

Lituya Bay
Glacier Bay
Haines
Skagway
Speed (Kts)
0
< .25
.25-1
1-1.5
1.5-2
2-2.5
2.5-3
3-4
> 4
Speed Unknown
Lynn Canal
Gustavus
OLGA-NEVA STRAITS
Pelican
Hoonah
Neva Strait
CHICHAGOF IS.
Juneau
Tenakee
Olga Strait
PERIL STRAIT
Seymour Canal
ADMIRALTY IS.
BARANOF IS.
Angoon
Sitka

Shaded Areas: See SELECTED CHANNELS

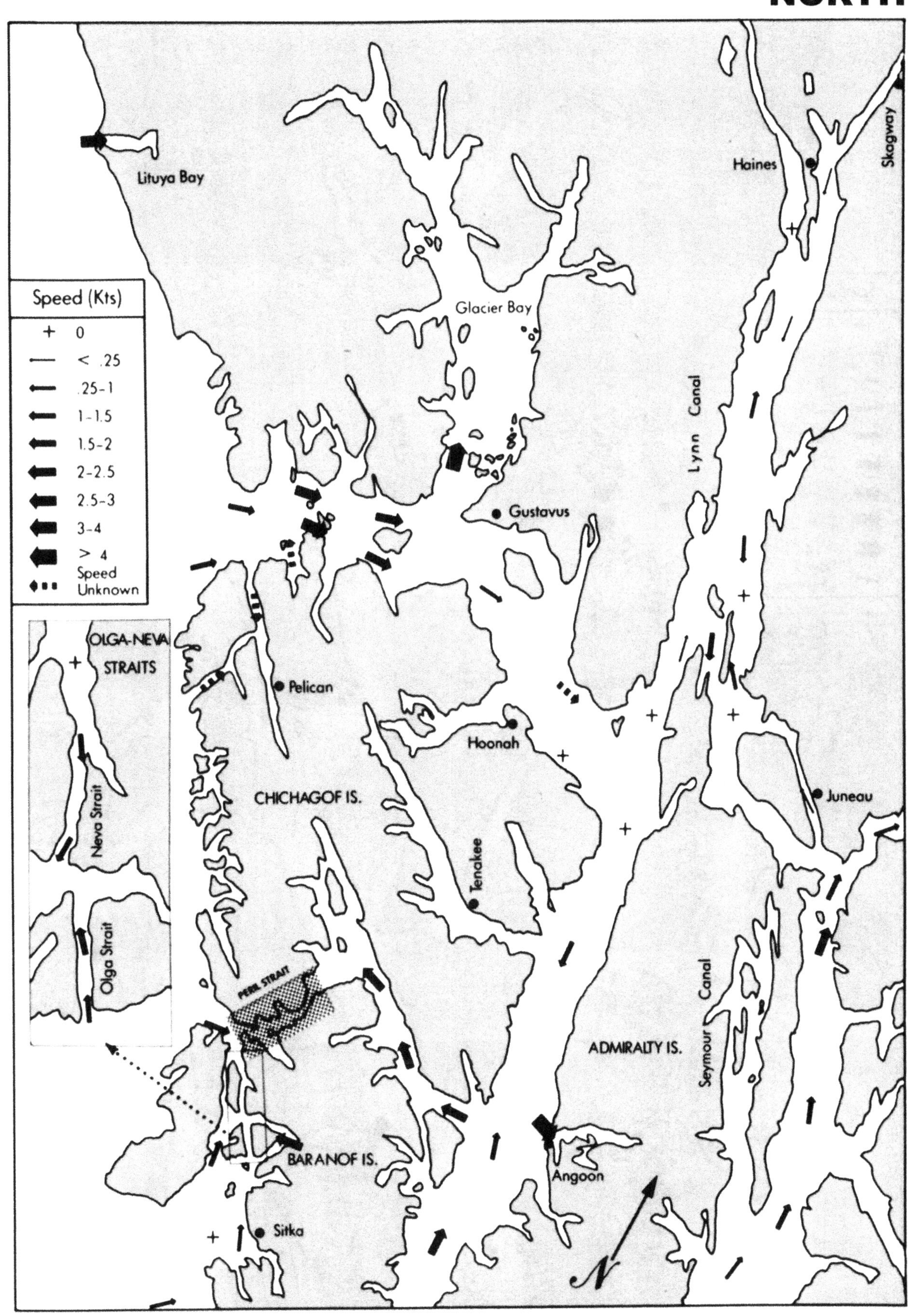

Lituya Bay
Glacier Bay
Haines
Skagway
Speed (Kts)
+ 0
< .25
.25-1
1-1.5
1.5-2
2-2.5
2.5-3
3-4
> 4
Speed Unknown
Lynn Canal
Gustavus
OLGA-NEVA STRAITS
Neva Strait
Olga Strait
Pelican
Hoonah
CHICHAGOF IS.
Juneau
Tenakee
PERIL STRAIT
Seymour Canal
ADMIRALTY IS.
BARANOF IS.
Angoon
Sitka
N

Shaded Areas: See SELECTED CHANNELS

NORTH

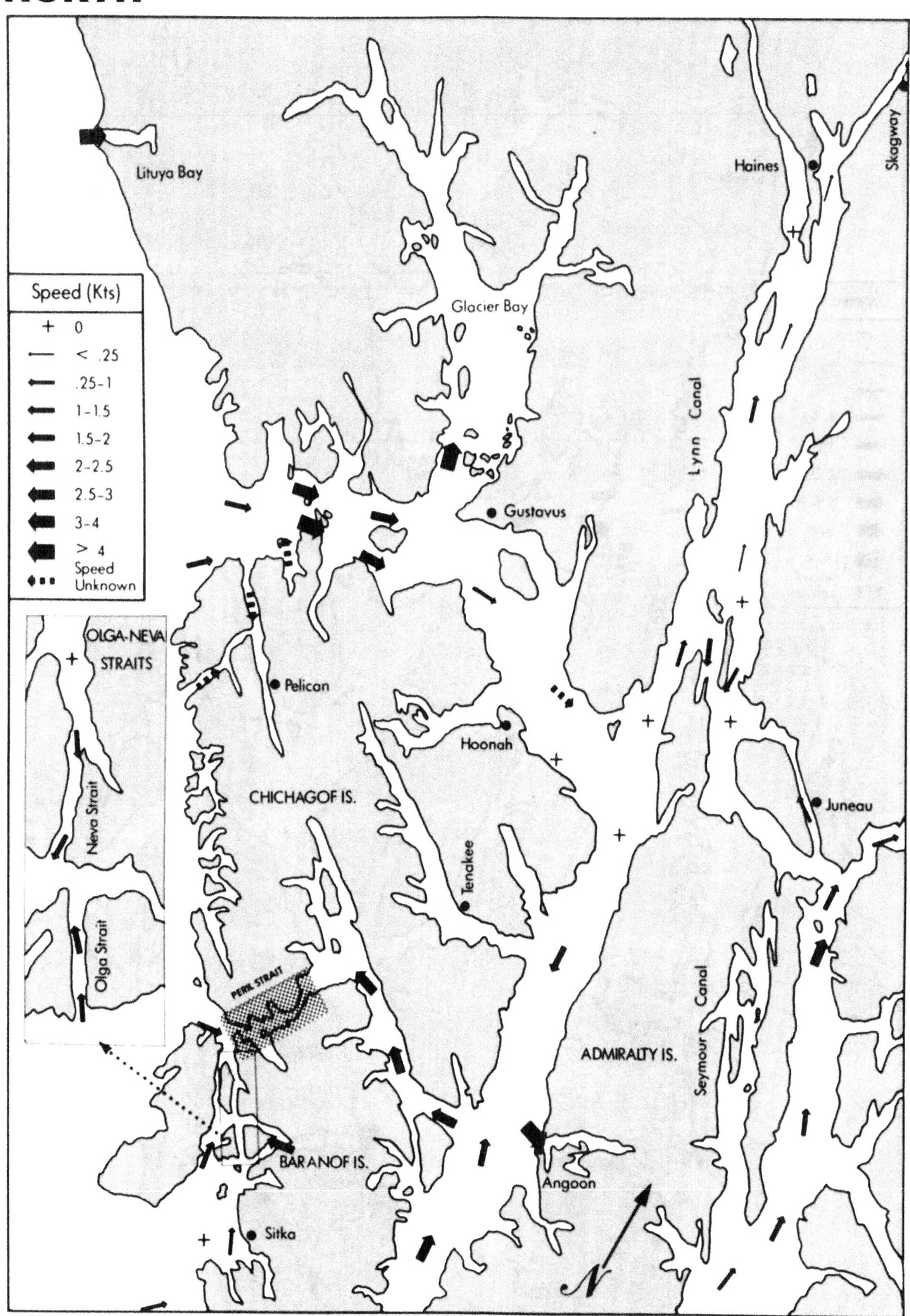

Lituya Bay
Haines
Skagway
Glacier Bay
Speed (Kts)
+ 0
< .25
.25-1
1-1.5
1.5-2
2-2.5
2.5-3
3-4
> 4
Speed Unknown
Lynn Canal
Gustavus
OLGA-NEVA STRAITS
Pelican
Hoonah
Neva Strait
CHICHAGOF IS.
Juneau
Tenakee
Olga Strait
PERIL STRAIT
Seymour Canal
ADMIRALTY IS.
BARANOF IS.
Angoon
Sitka
N

Shaded Areas: See SELECTED CHANNELS

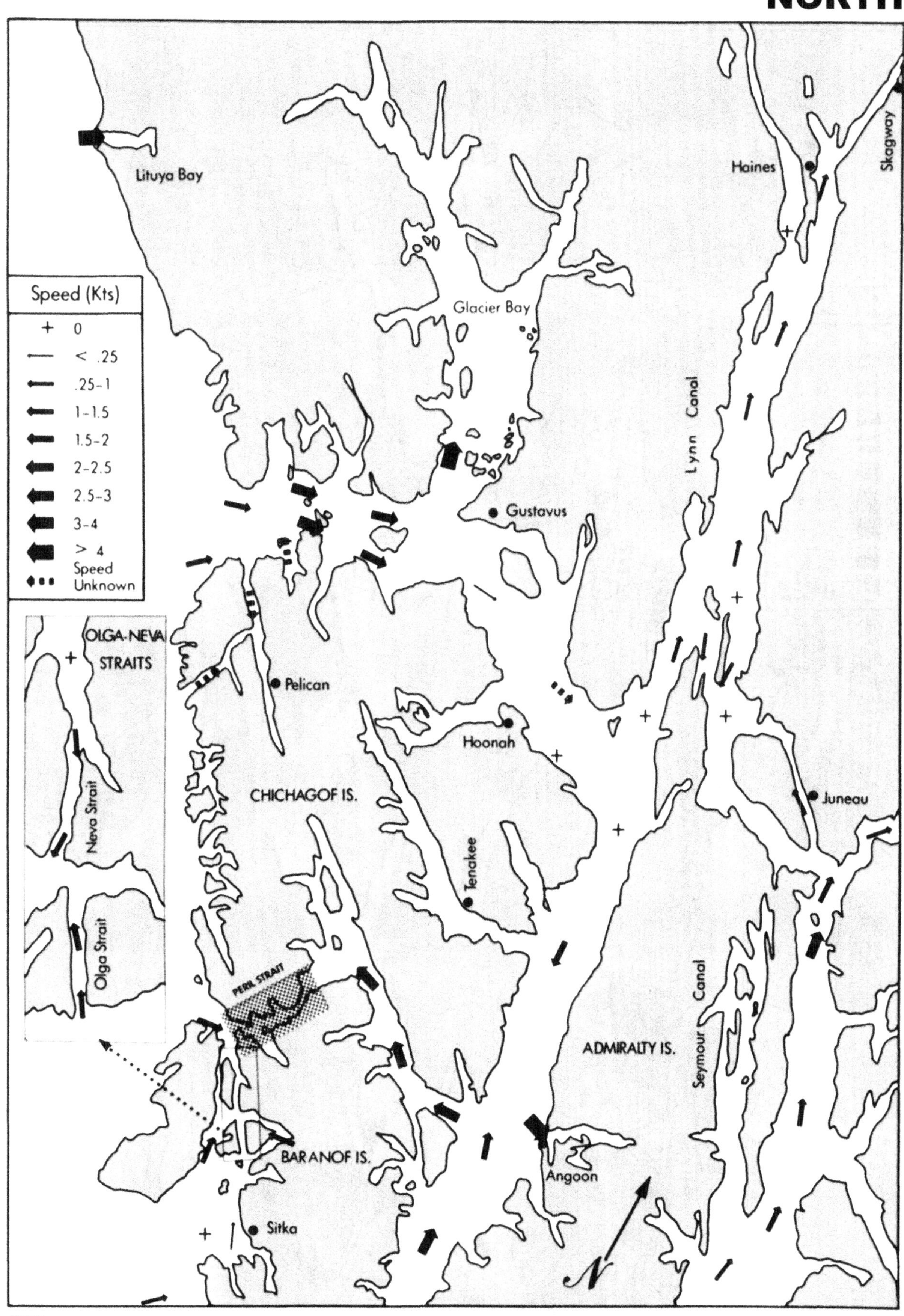
Lituya Bay
Haines
Skagway
Speed (Kts)
+ 0
< .25
.25-1
1-1.5
1.5-2
2-2.5
2.5-3
3-4
> 4
Speed Unknown
Glacier Bay
Lynn Canal
Gustavus
OLGA-NEVA STRAITS
Pelican
Hoonah
Juneau
CHICHAGOF IS.
Neva Strait
Tenakee
Olga Strait
PERIL STRAIT
Seymour Canal
ADMIRALTY IS.
BARANOF IS.
Angoon
Sitka
N

Shaded Areas: See SELECTED CHANNELS

NORTH

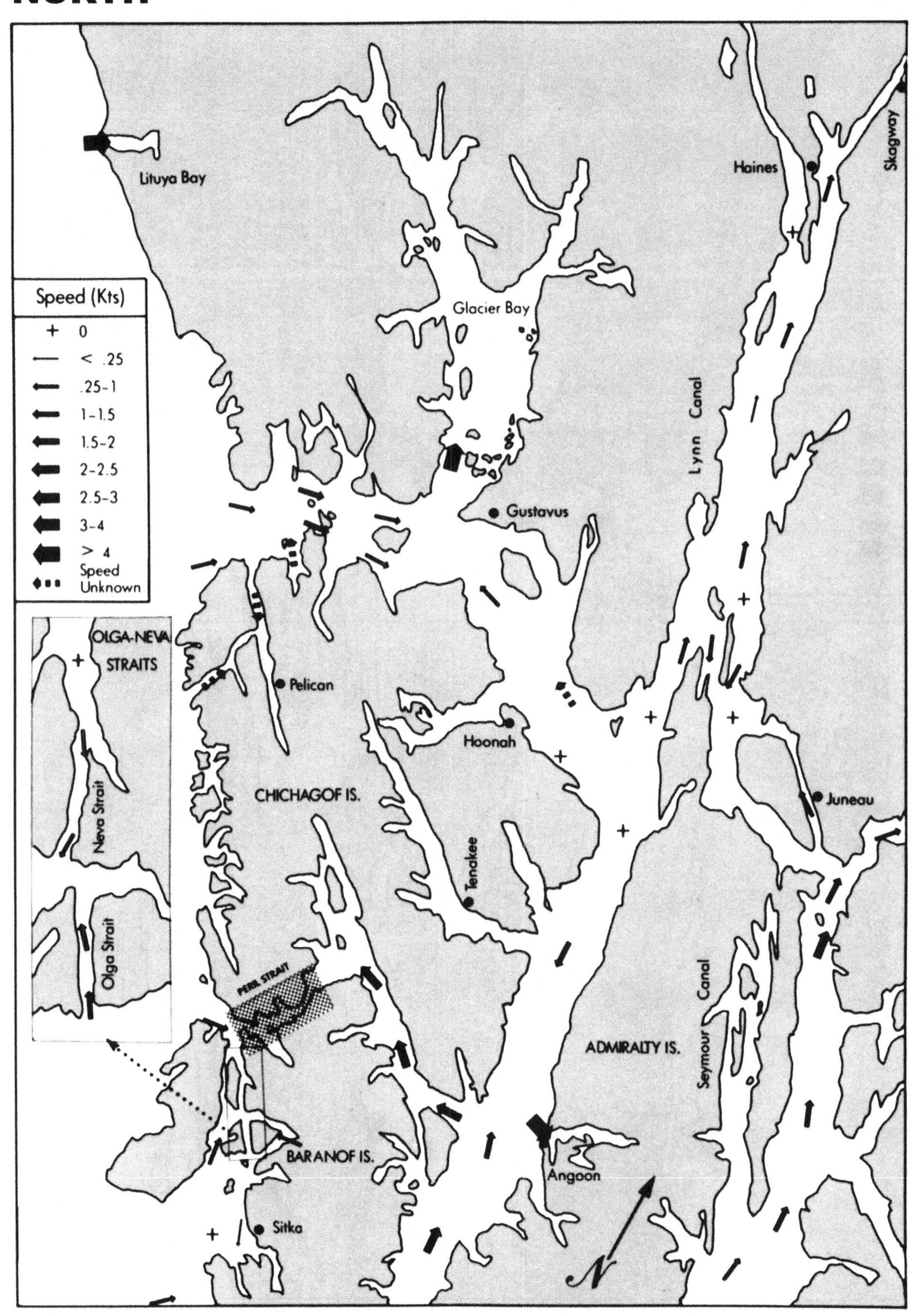

Lituya Bay
Glacier Bay
Haines
Skagway
Lynn Canal
Gustavus
Pelican
Hoonah
Juneau
CHICHAGOF IS.
Tenakee
OLGA-NEVA STRAITS
Neva Strait
Olga Strait
PERIL STRAIT
BARANOF IS.
Sitka
Angoon
ADMIRALTY IS.
Seymour Canal
Speed (Kts)
+ 0
< .25
.25-1
1-1.5
1.5-2
2-2.5
2.5-3
3-4
> 4
Speed Unknown

Shaded Areas: See SELECTED CHANNELS

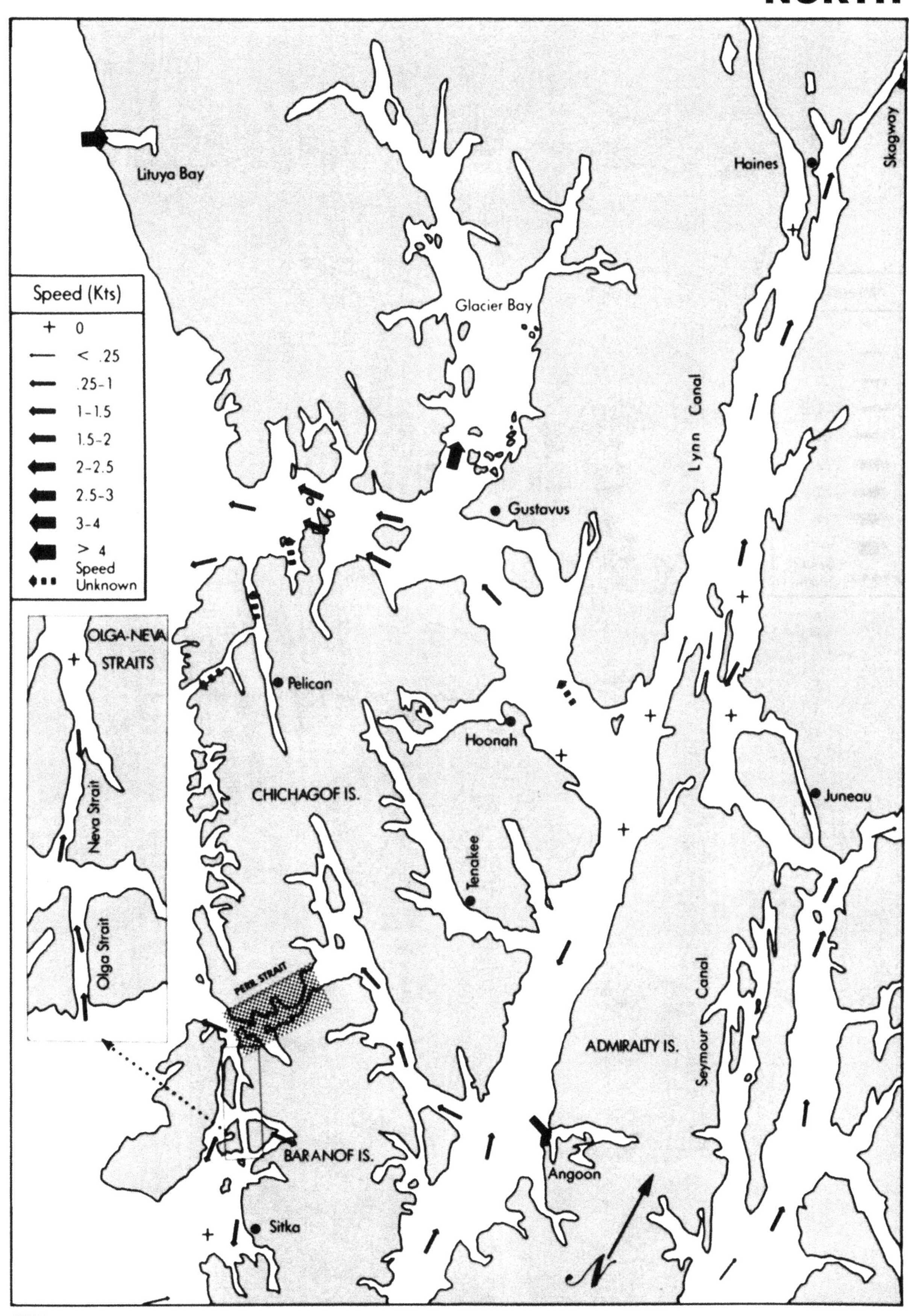
Lituya Bay
Speed (Kts)
+ 0
< .25
.25-1
1-1.5
1.5-2
2-2.5
2.5-3
3-4
> 4
Speed Unknown
Glacier Bay
Haines
Skagway
Lynn Canal
Gustavus
OLGA-NEVA STRAITS
Neva Strait
Olga Strait
Pelican
Hoonah
CHICHAGOF IS.
Juneau
Tenakee
PERIL STRAIT
ADMIRALTY IS.
Seymour Canal
BARANOF IS.
Angoon
Sitka
N

Shaded Areas: See SELECTED CHANNELS

NORTH

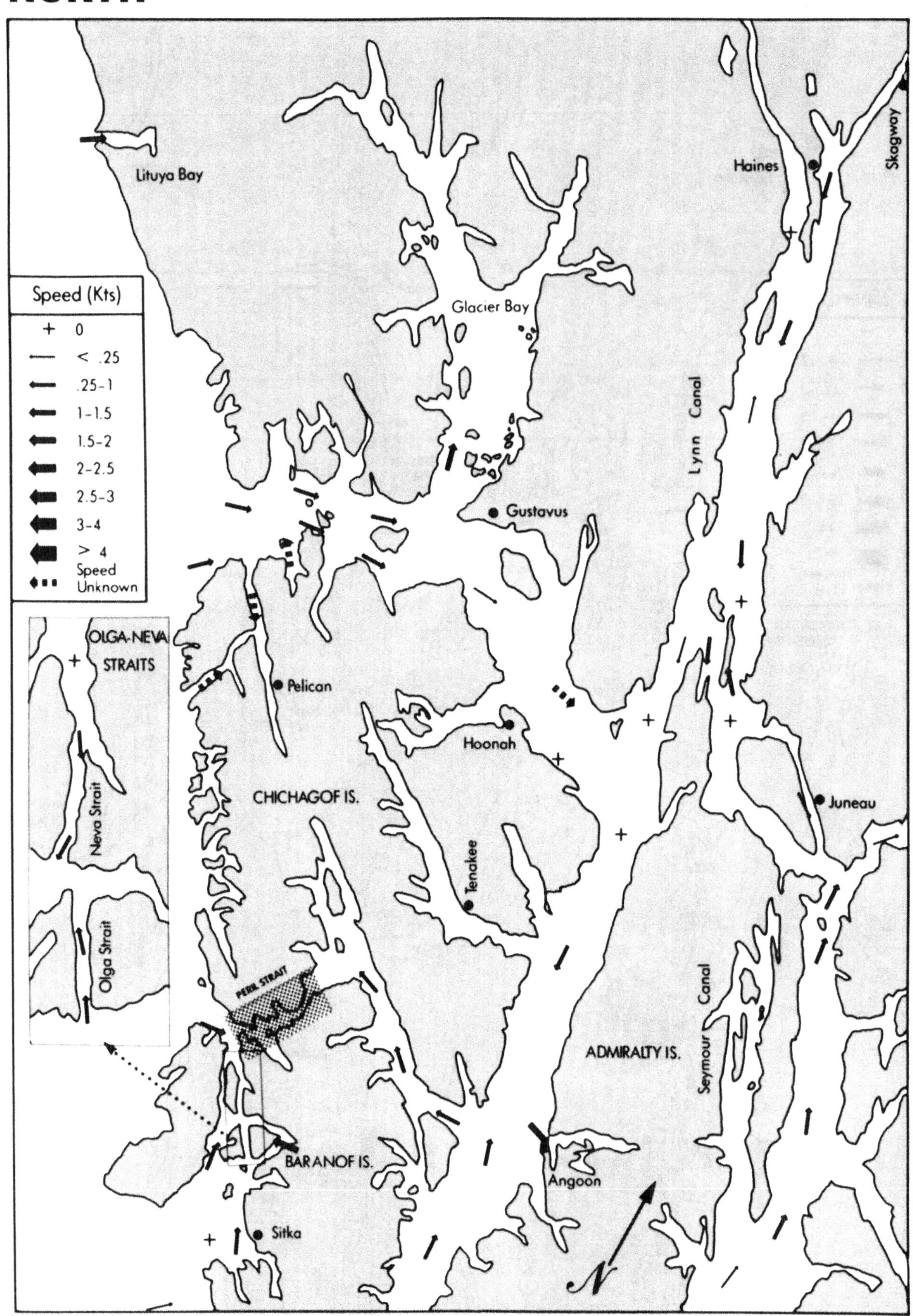

Lituya Bay
Haines
Skagway
Glacier Bay
Speed (Kts)
+ 0
< .25
.25-1
1-1.5
1.5-2
2-2.5
2.5-3
3-4
> 4
Speed Unknown
Lynn Canal
Gustavus
OLGA-NEVA STRAITS
Pelican
Hoonah
Neva Strait
CHICHAGOF IS.
Juneau
Tenakee
Olga Strait
PERIL STRAIT
Seymour Canal
ADMIRALTY IS.
BARANOF IS.
Angoon
Sitka
N

Shaded Areas: See SELECTED CHANNELS

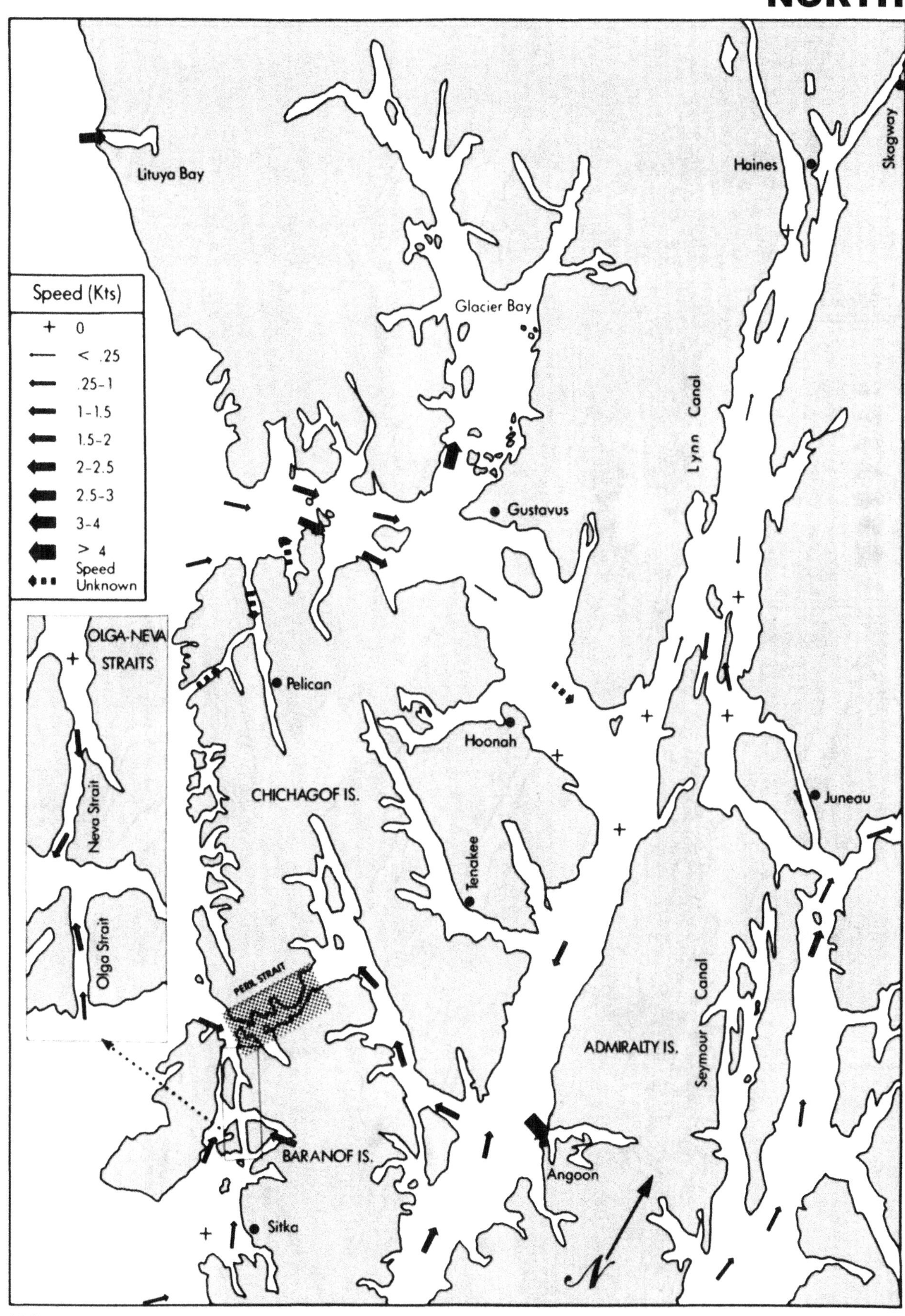
Lituya Bay
Glacier Bay
Haines
Skogway
Speed (Kts)
+ 0
< .25
.25-1
1-1.5
1.5-2
2-2.5
2.5-3
3-4
> 4
Speed Unknown
Lynn Canal
Gustavus
OLGA-NEVA STRAITS
Pelican
Hoonah
Juneau
CHICHAGOF IS.
Neva Strait
Tenakee
Olga Strait
PERIL STRAIT
Seymour Canal
ADMIRALTY IS.
BARANOF IS.
Angoon
Sitka
N

Shaded Areas: See SELECTED CHANNELS

NORTH

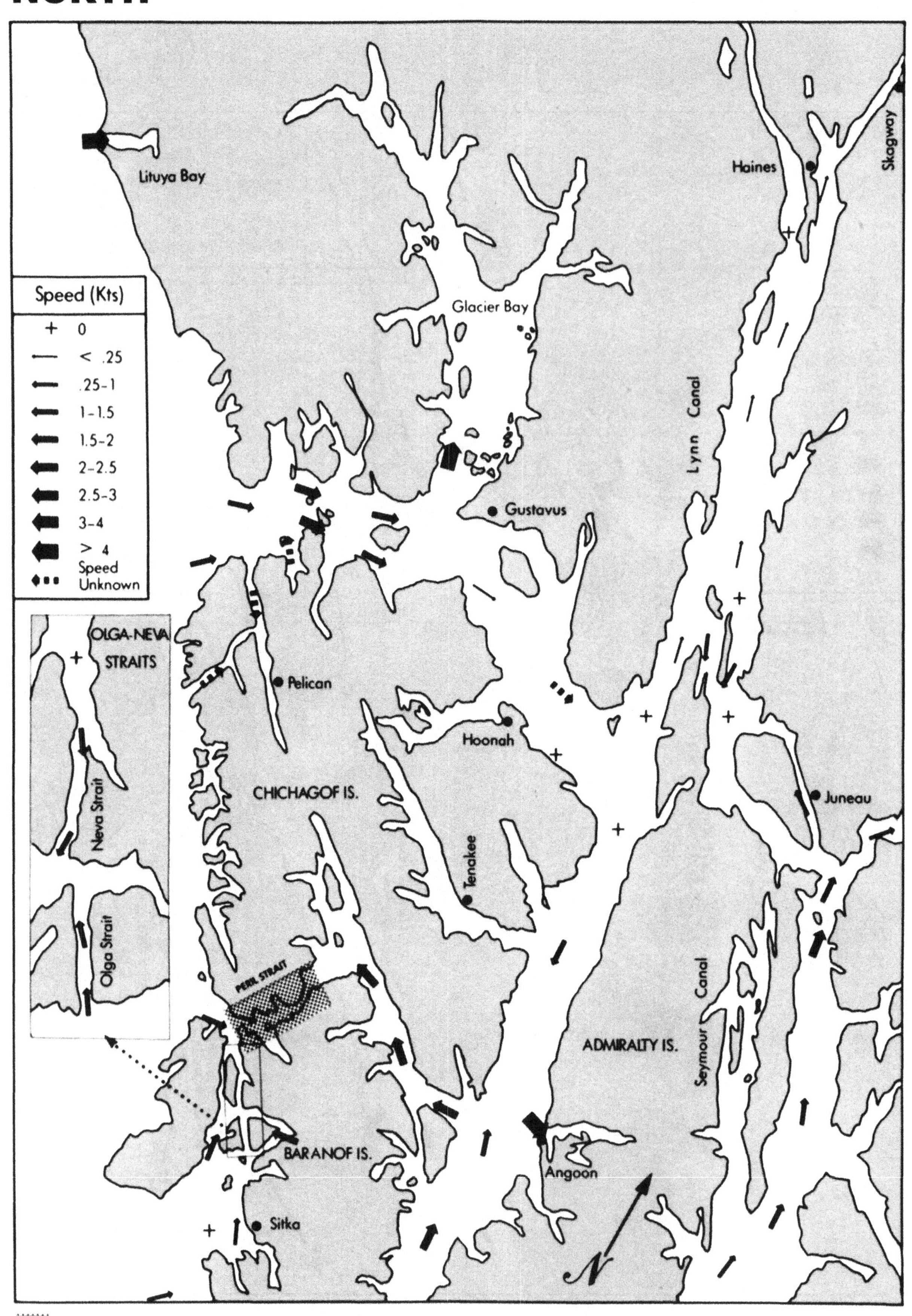

Lituya Bay
Haines
Skagway
Speed (Kts)
0
< .25
.25-1
1-1.5
1.5-2
2-2.5
2.5-3
3-4
> 4
Speed Unknown
Glacier Bay
Lynn Canal
Gustavus
OLGA-NEVA STRAITS
Pelican
Hoonah
CHICHAGOF IS.
Juneau
Neva Strait
Tenakee
Olga Strait
PERIL STRAIT
Seymour Canal
ADMIRALTY IS.
BARANOF IS.
Angoon
Sitka
N

Shaded Areas: See SELECTED CHANNELS

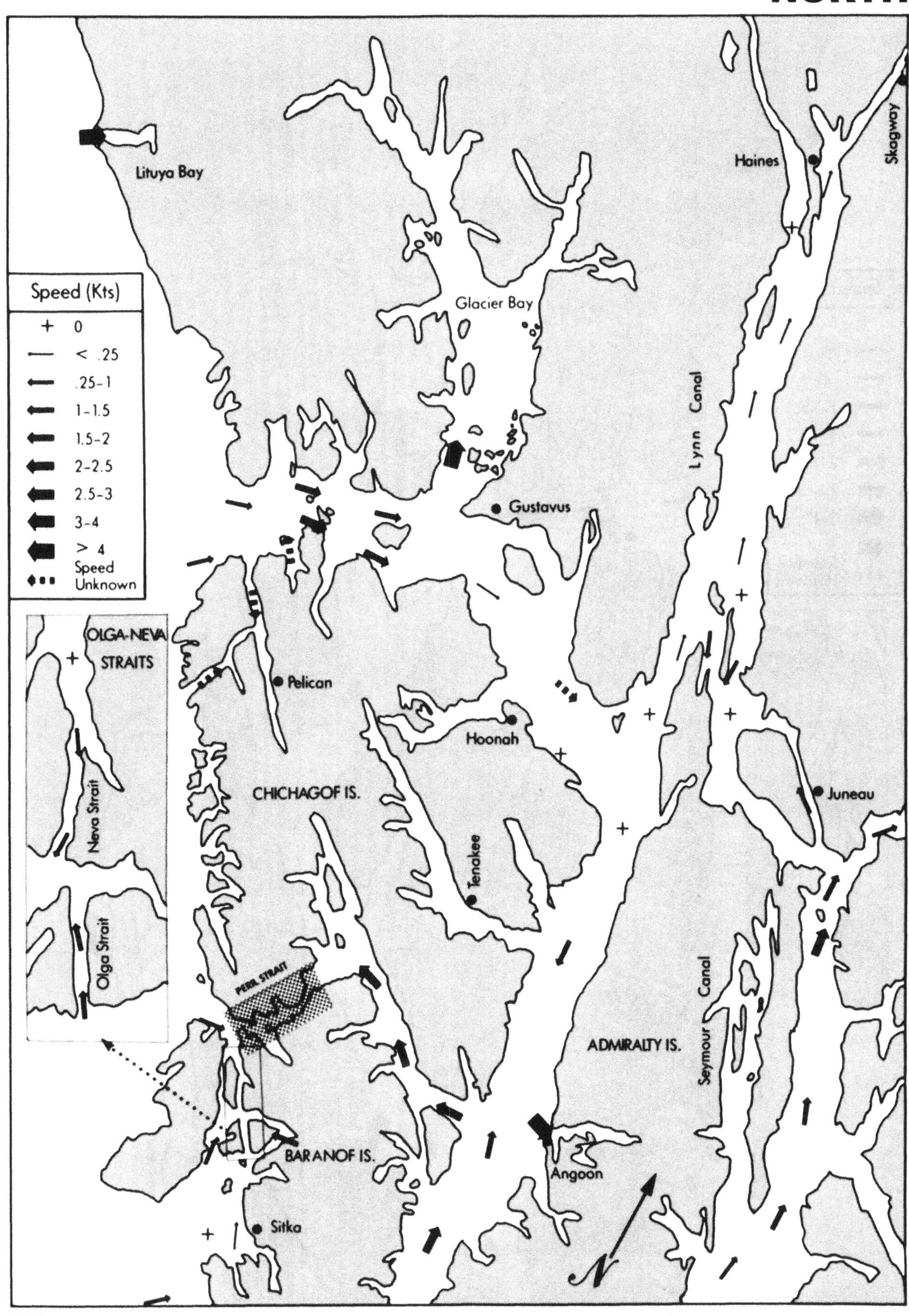

Lituya Bay
Glacier Bay
Haines
Skagway
Lynn Canal
Gustavus
Pelican
Hoonah
Juneau
CHICHAGOF IS.
Tenakee
ADMIRALTY IS.
Seymour Canal
PERIL STRAIT
BARANOF IS.
Angoon
Sitka
Speed (Kts)
0
< .25
.25-1
1-1.5
1.5-2
2-2.5
2.5-3
3-4
> 4
Speed Unknown
OLGA-NEVA STRAITS
Neva Strait
Olga Strait

Shaded Areas: See SELECTED CHANNELS

NORTH

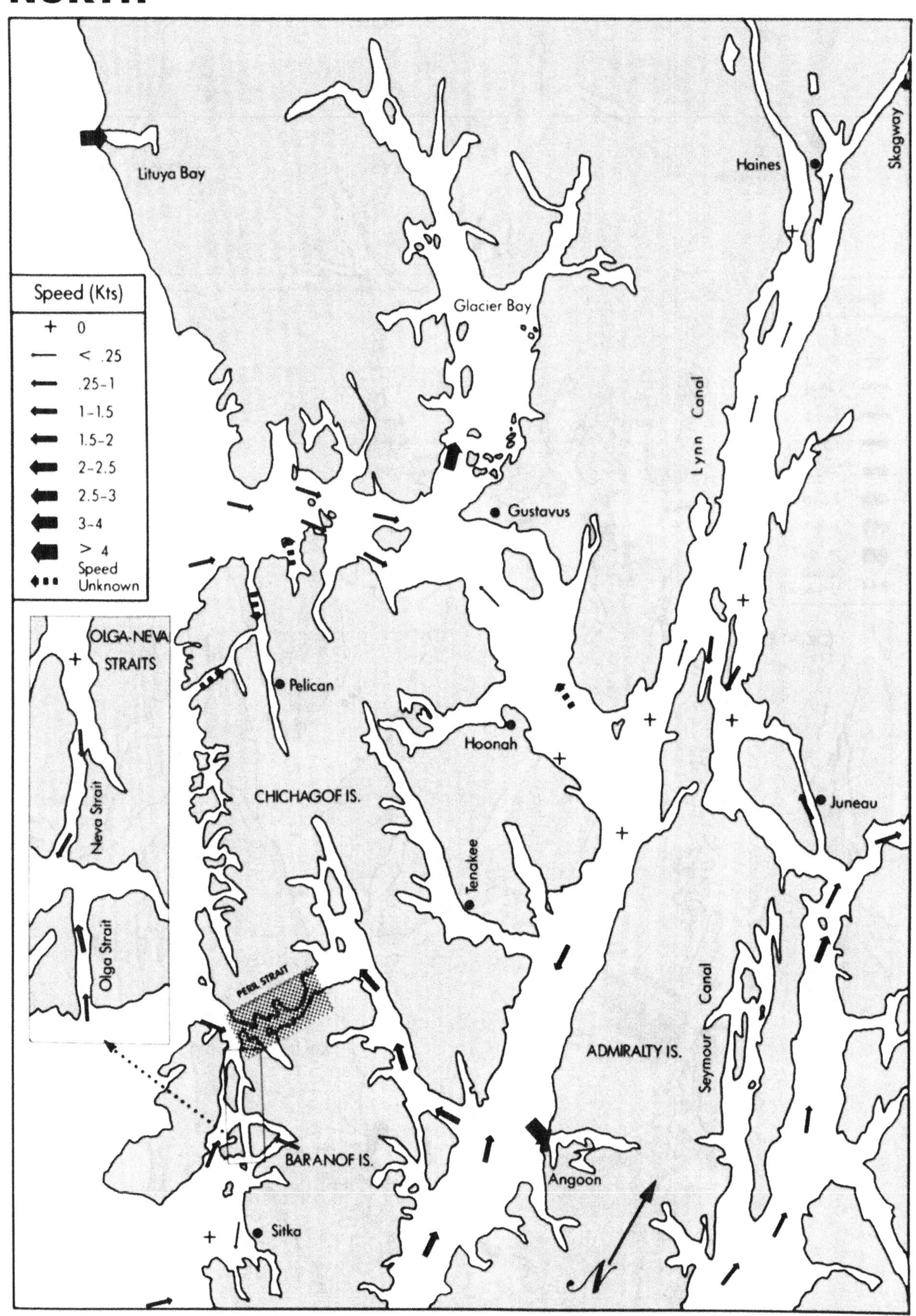

Lituya Bay
Haines
Skagway
Glacier Bay
Speed (Kts)
+ 0
< .25
.25-1
1-1.5
1.5-2
2-2.5
2.5-3
3-4
> 4
Speed Unknown
Lynn Canal
Gustavus
OLGA-NEVA STRAITS
Neva Strait
Olga Strait
Pelican
Hoonah
Juneau
CHICHAGOF IS.
Tenakee
PERIL STRAIT
Seymour Canal
ADMIRALTY IS.
BARANOF IS.
Angoon
Sitka
N

Shaded Areas: See SELECTED CHANNELS

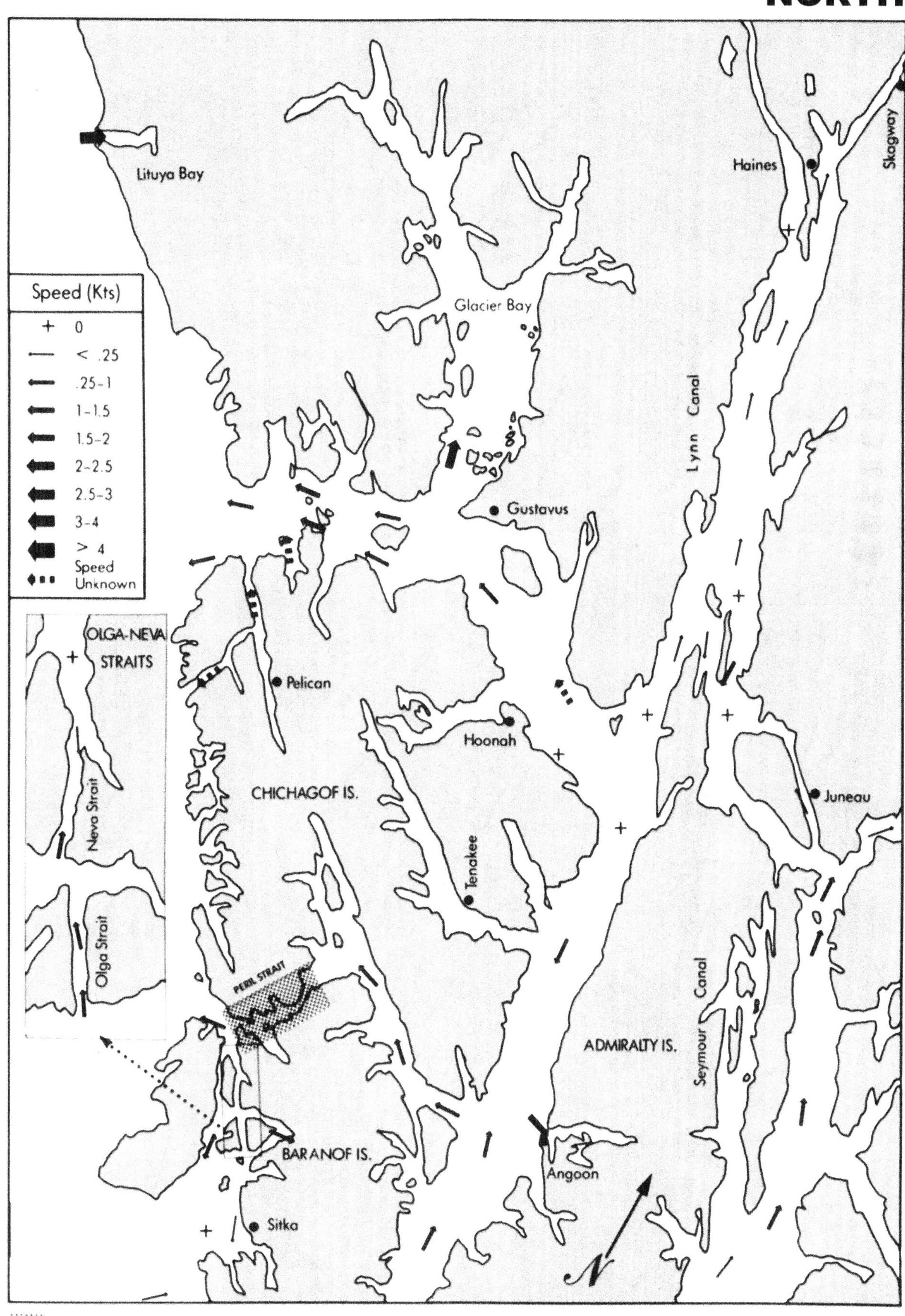
Lituya Bay
Haines
Skagway
Glacier Bay
Speed (Kts)
+ 0
< .25
.25-1
1-1.5
1.5-2
2-2.5
2.5-3
3-4
> 4
Speed Unknown
Lynn Canal
Gustavus
OLGA-NEVA STRAITS
Pelican
Hoonah
Juneau
Neva Strait
Olga Strait
CHICHAGOF IS.
Tenakee
PERIL STRAIT
Seymour Canal
ADMIRALTY IS.
BARANOF IS.
Angoon
Sitka
N

Shaded Areas: See SELECTED CHANNELS

NORTH

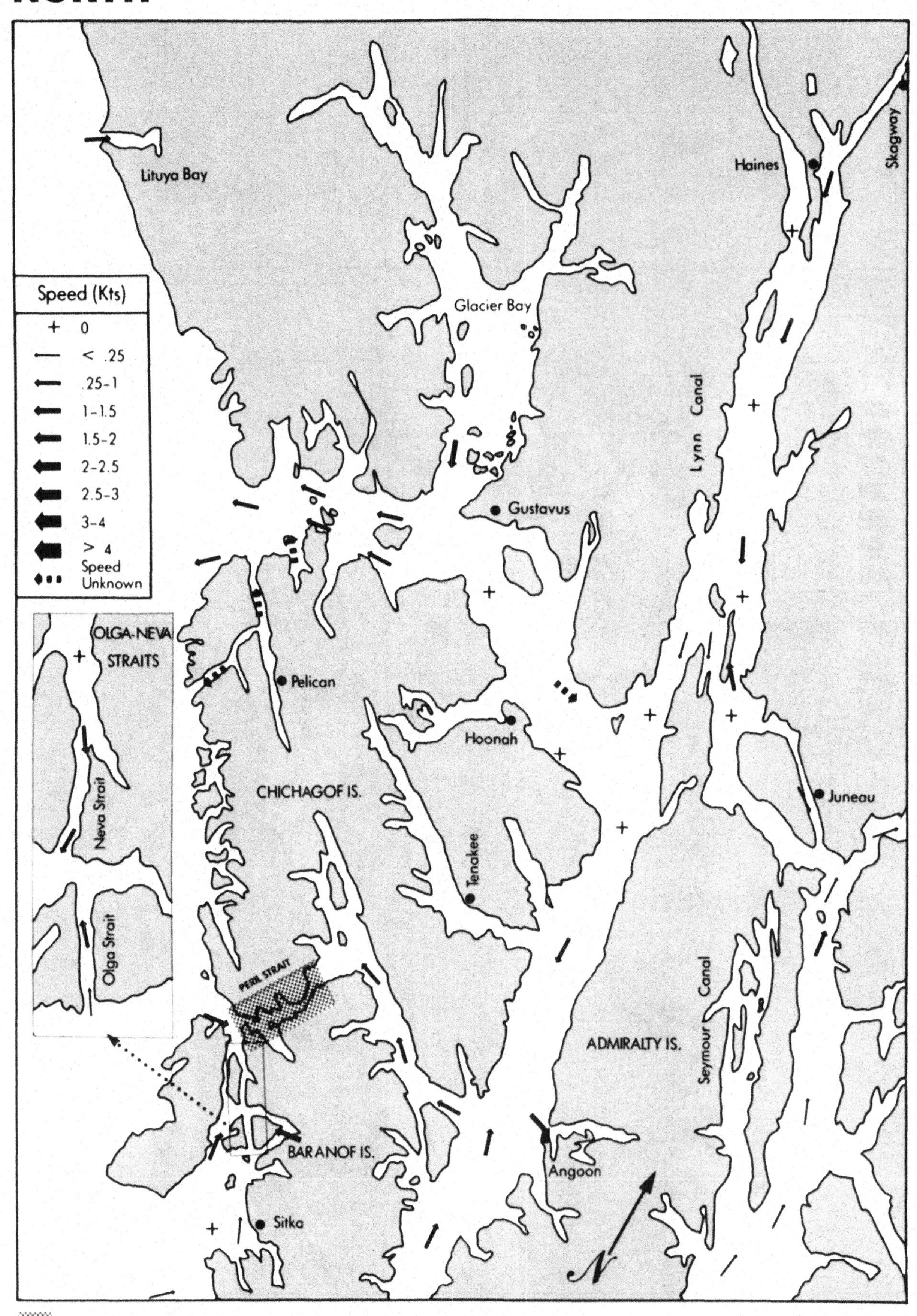

Lituya Bay
Glacier Bay
Haines
Skagway
Speed (Kts)
+ 0
< .25
.25-1
1-1.5
1.5-2
2-2.5
2.5-3
3-4
> 4
Speed Unknown
Lynn Canal
Gustavus
OLGA-NEVA STRAITS
Neva Strait
Olga Strait
Pelican
Hoonah
CHICHAGOF IS.
Juneau
Tenakee
PERIL STRAIT
ADMIRALTY IS.
Seymour Canal
BARANOF IS.
Angoon
Sitka
N

Shaded Areas: See SELECTED CHANNELS

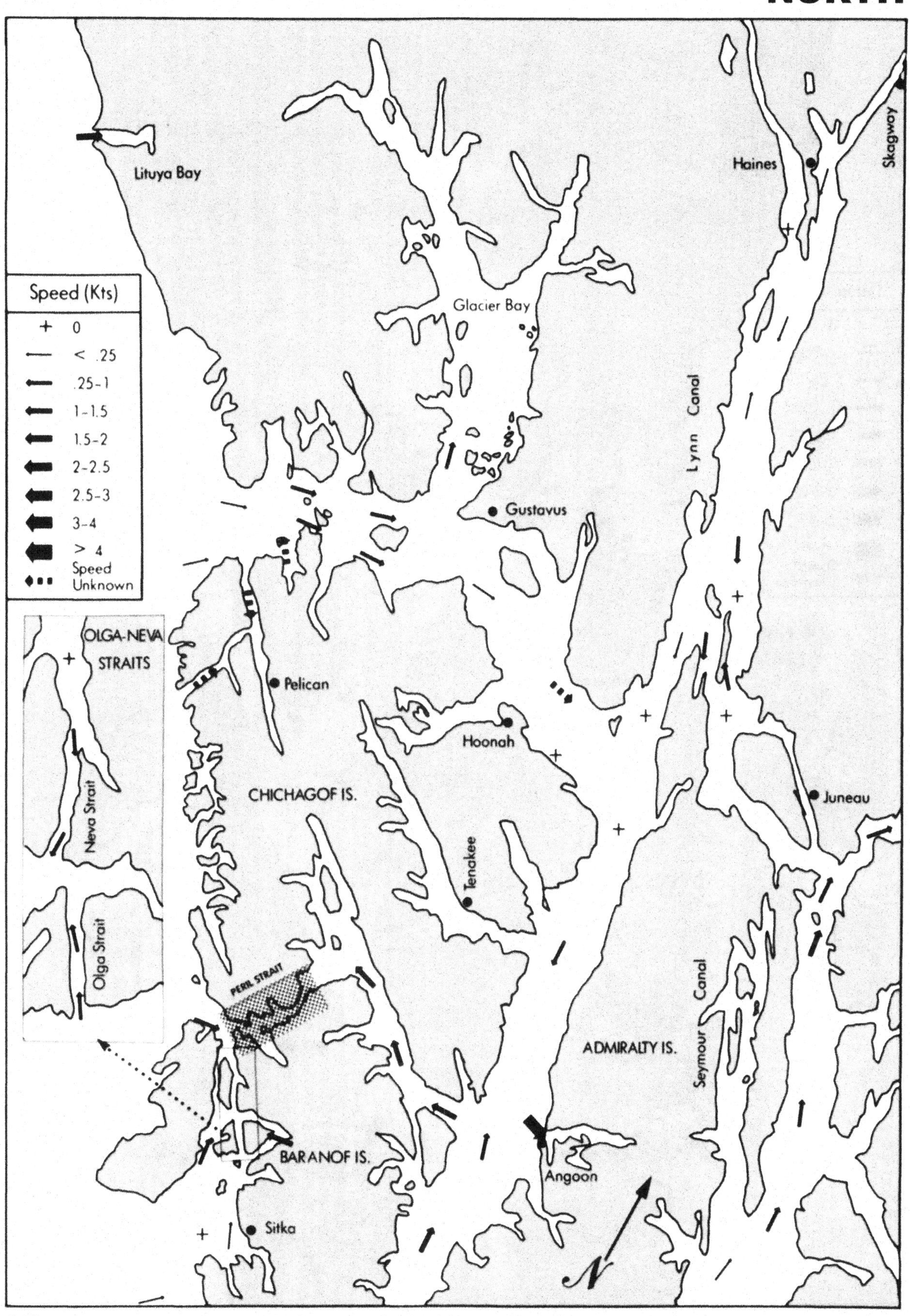
Lituya Bay
Glacier Bay
Haines
Skagway
Lynn Canal
Gustavus
Pelican
Hoonah
Juneau
CHICHAGOF IS.
Tenakee
OLGA-NEVA
STRAITS
Neva Strait
Olga Strait
PERIL STRAIT
Seymour Canal
ADMIRALTY IS.
BARANOF IS.
Angoon
Sitka
Speed (Kts)
0
< .25
.25-1
1-1.5
1.5-2
2-2.5
2.5-3
3-4
> 4
Speed
Unknown

Shaded Areas: See SELECTED CHANNELS

NORTH

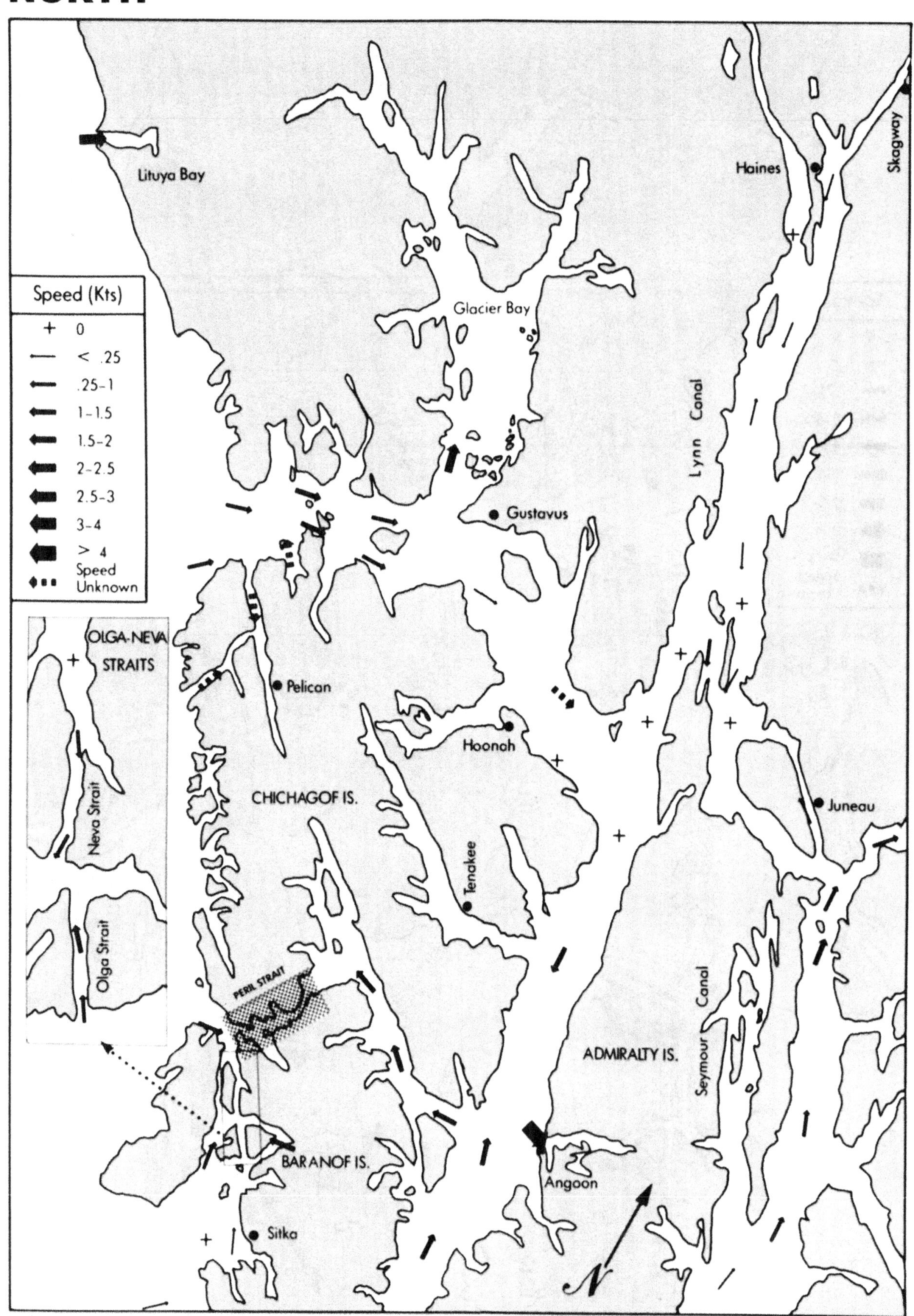

Speed (Kts)
+ 0
< .25
.25-1
1-1.5
1.5-2
2-2.5
2.5-3
3-4
> 4
Speed Unknown
Lituya Bay
Glacier Bay
Haines
Skagway
Lynn Canal
Gustavus
Pelican
Hoonah
Juneau
CHICHAGOF IS.
Tenakee
OLGA-NEVA STRAITS
Neva Strait
Olga Strait
PERIL STRAIT
ADMIRALTY IS.
Seymour Canal
BARANOF IS.
Angoon
Sitka
N

Shaded Areas: See SELECTED CHANNELS

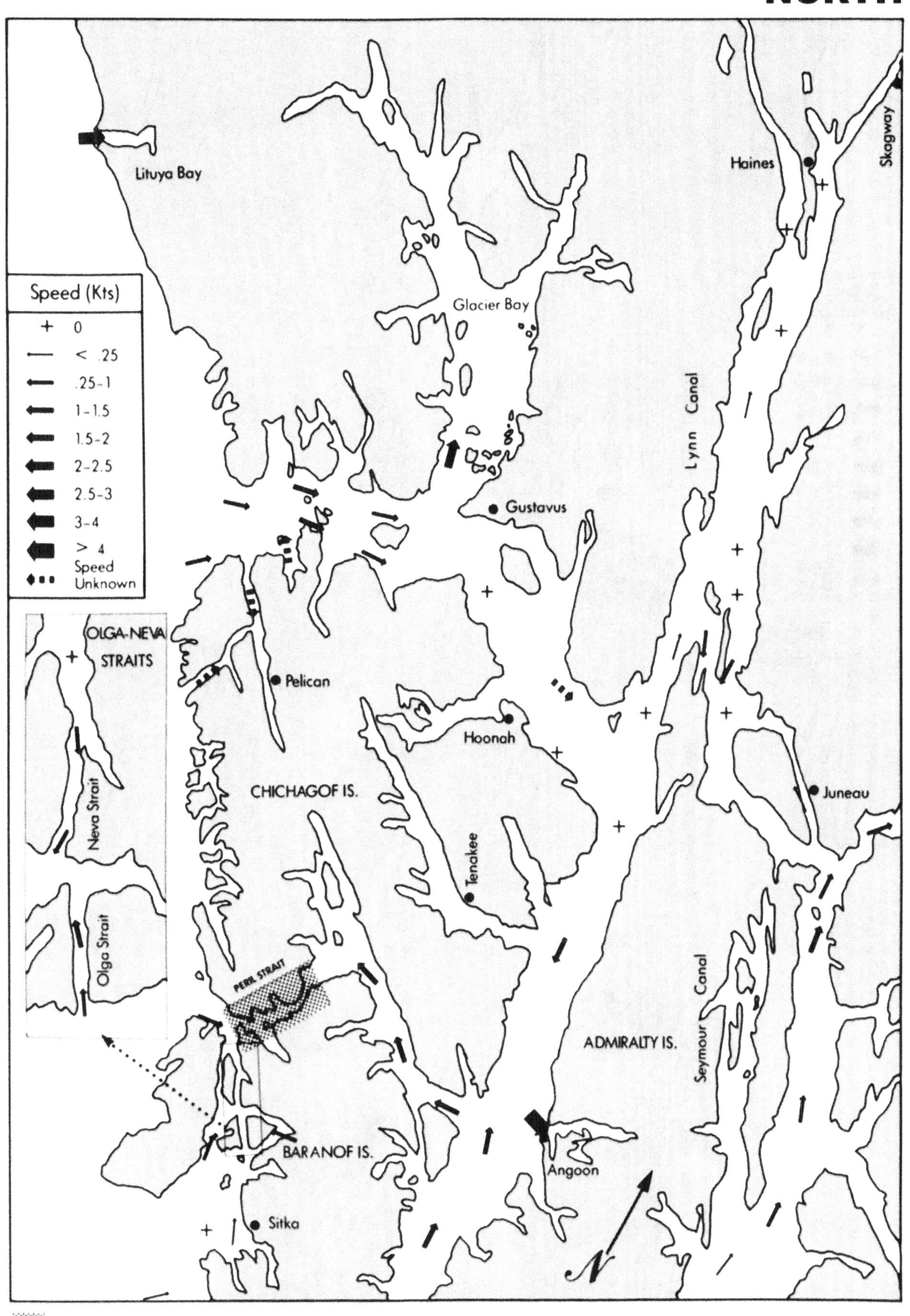
Lituya Bay
Haines
Skagway
Speed (Kts)
+ 0
< .25
.25-1
1-1.5
1.5-2
2-2.5
2.5-3
3-4
> 4
Speed Unknown
Glacier Bay
Lynn Canal
Gustavus
OLGA-NEVA STRAITS
Pelican
Hoonah
Juneau
Neva Strait
CHICHAGOF IS.
Tenakee
Olga Strait
PERIL STRAIT
Seymour Canal
ADMIRALTY IS.
BARANOF IS.
Angoon
Sitka
N

Shaded Areas: See SELECTED CHANNELS

NORTH

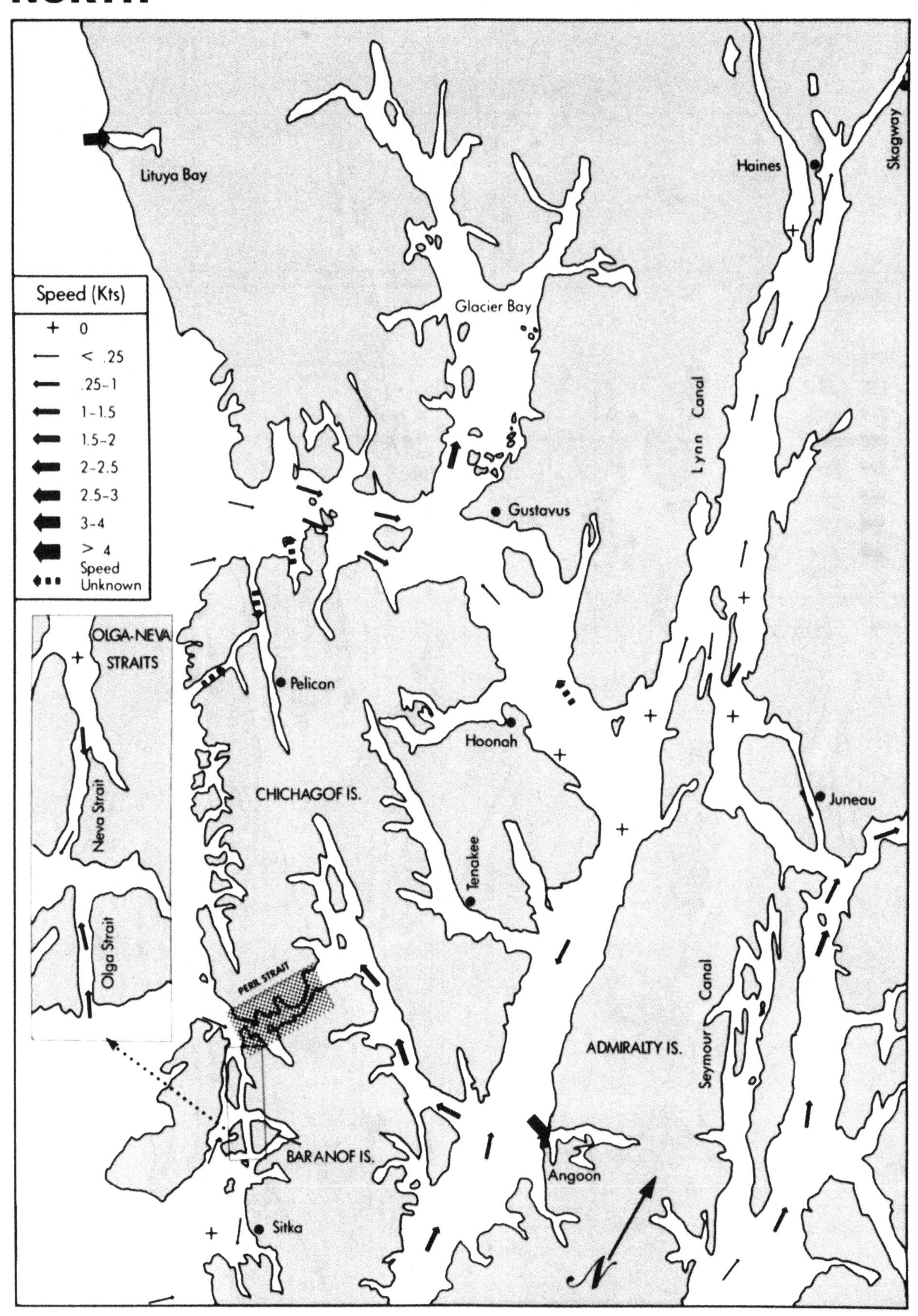

Lituya Bay
Haines
Skagway
Speed (Kts)
+ 0
< .25
.25-1
1-1.5
1.5-2
2-2.5
2.5-3
3-4
> 4
Speed Unknown
Glacier Bay
Lynn Canal
Gustavus
OLGA-NEVA STRAITS
Pelican
Hoonah
Neva Strait
CHICHAGOF IS.
Juneau
Tenakee
Olga Strait
PERIL STRAIT
Seymour Canal
ADMIRALTY IS.
BARANOF IS.
Angoon
Sitka
N

Shaded Areas: See SELECTED CHANNELS

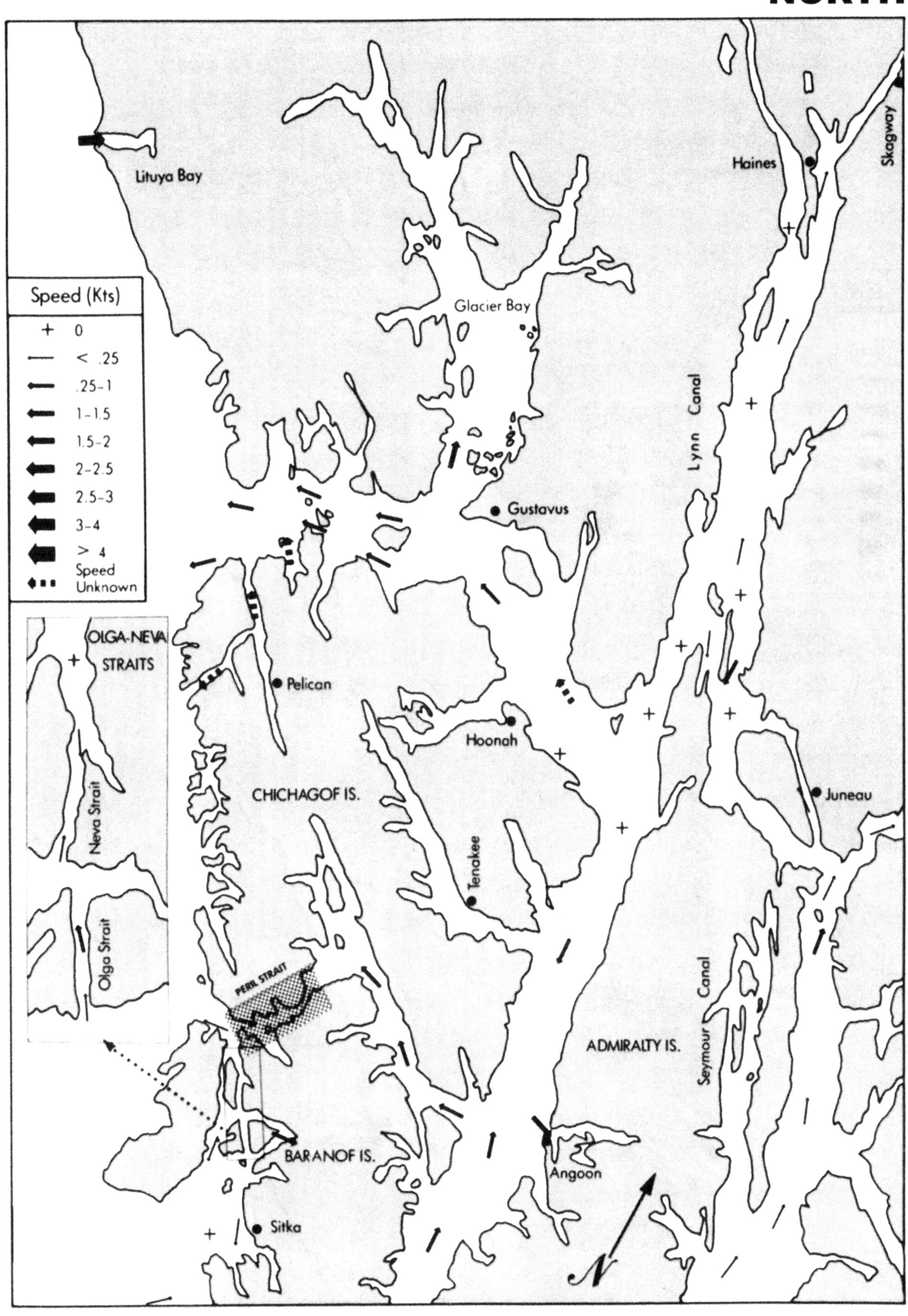
Lituya Bay
Haines
Skagway
Glacier Bay
Speed (Kts)
+ 0
< .25
.25-1
1-1.5
1.5-2
2-2.5
2.5-3
3-4
> 4
Speed Unknown
Gustavus
Lynn Canal
OLGA-NEVA STRAITS
Pelican
Hoonah
Juneau
CHICHAGOF IS.
Neva Strait
Tenakee
Olga Strait
PERIL STRAIT
Seymour Canal
ADMIRALTY IS.
BARANOF IS.
Angoon
Sitka
N

Shaded Areas: See SELECTED CHANNELS

NORTH

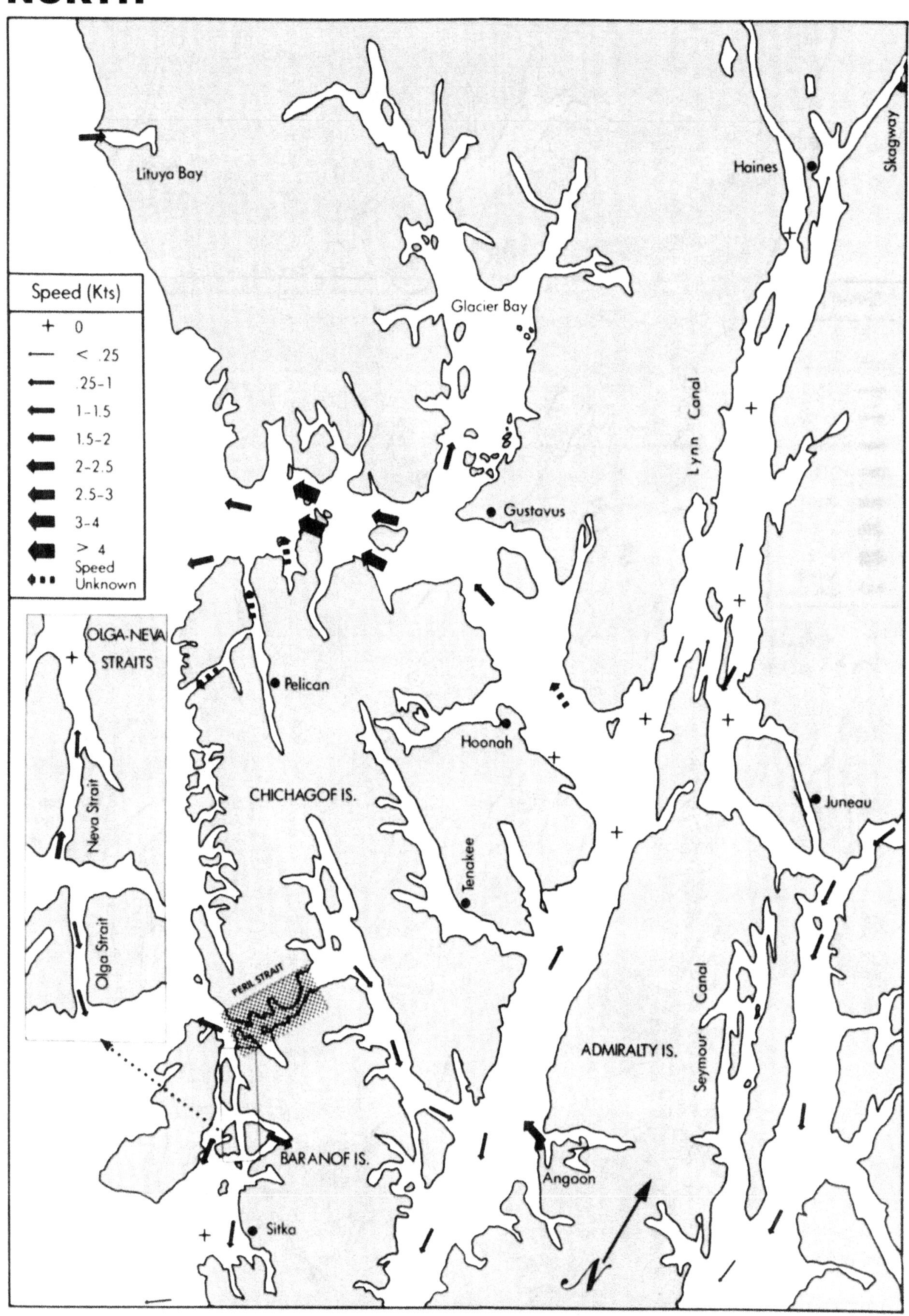

Lituya Bay
Haines
Skagway
Glacier Bay
Speed (Kts)
+ 0
< .25
.25-1
1-1.5
1.5-2
2-2.5
2.5-3
3-4
> 4
Speed Unknown
Lynn Canal
Gustavus
OLGA-NEVA STRAITS
Pelican
Hoonah
Juneau
Neva Strait
CHICHAGOF IS.
Tenakee
Olga Strait
PERIL STRAIT
Seymour Canal
ADMIRALTY IS.
BARANOF IS.
Angoon
Sitka
N

Shaded Areas: See SELECTED CHANNELS

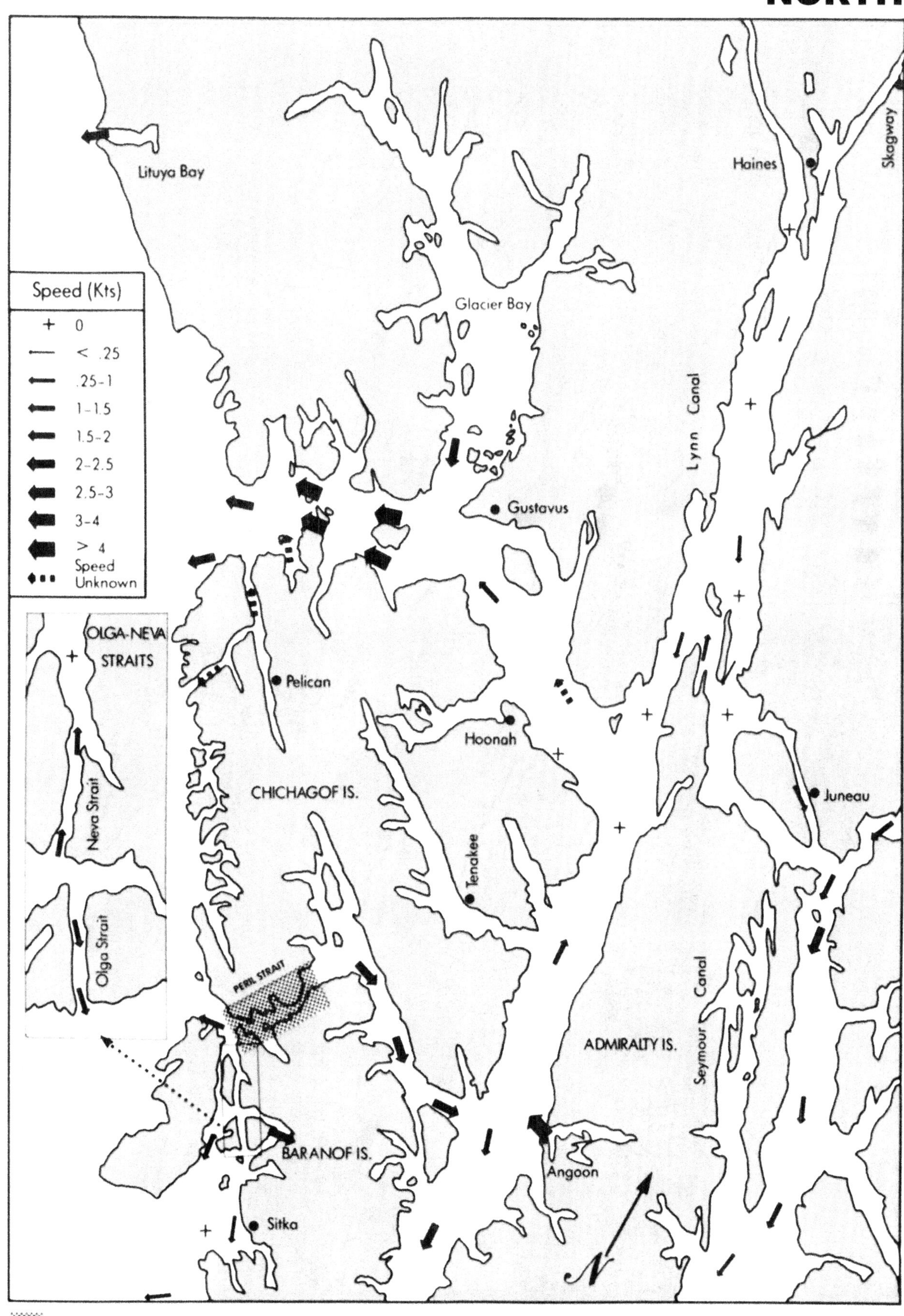
Lituya Bay
Glacier Bay
Haines
Skagway
Speed (Kts)
0
< .25
.25-1
1-1.5
1.5-2
2-2.5
2.5-3
3-4
> 4
Speed Unknown
Lynn Canal
Gustavus
OLGA-NEVA STRAITS
Pelican
Hoonah
Neva Strait
Olga Strait
CHICHAGOF IS.
Juneau
Tenakee
PERIL STRAIT
Seymour Canal
ADMIRALTY IS.
BARANOF IS.
Angoon
Sitka
N

Shaded Areas: See SELECTED CHANNELS

NORTH

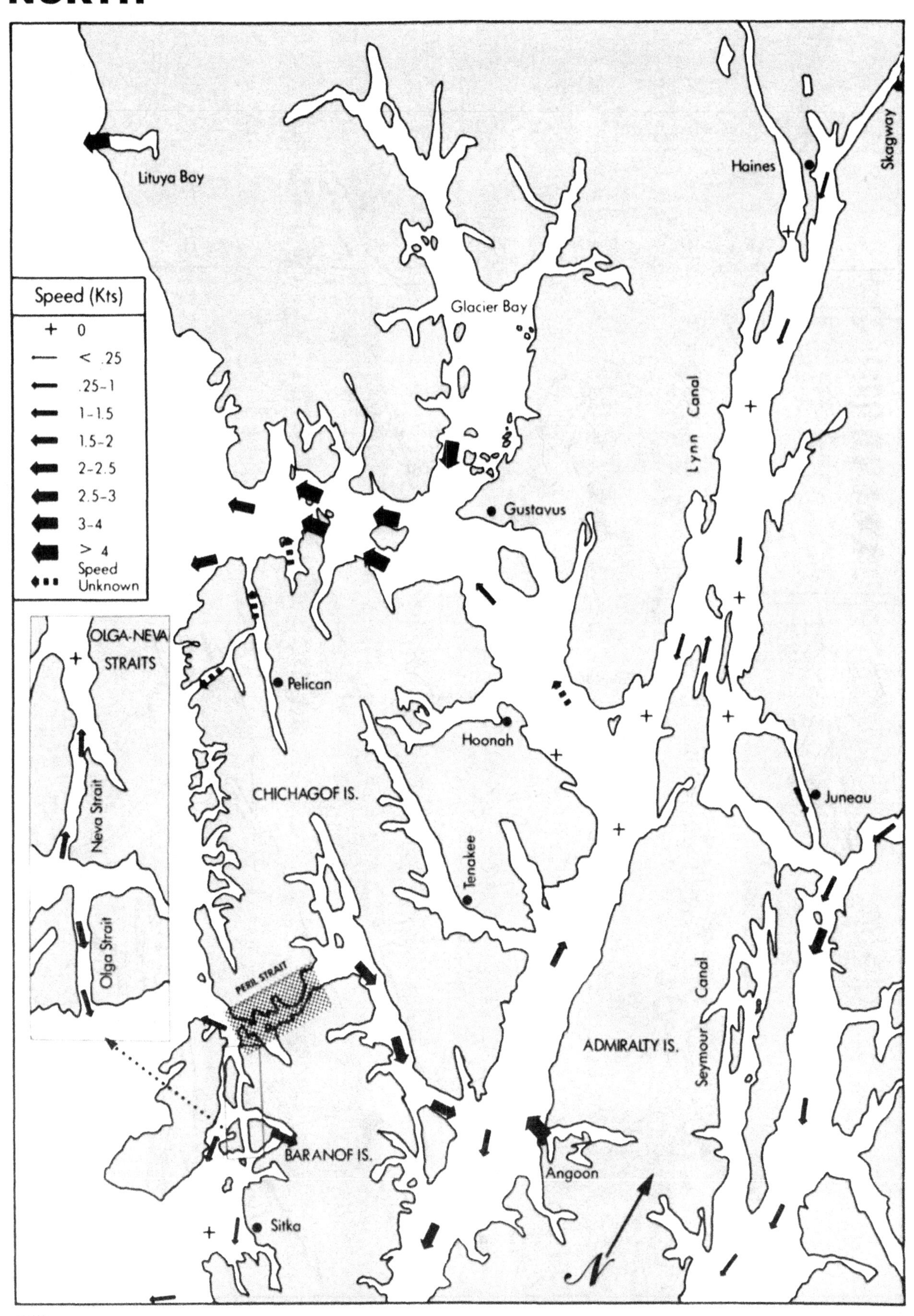

Lituya Bay
Haines
Skagway
Glacier Bay
Speed (Kts)
+ 0
< .25
.25-1
1-1.5
1.5-2
2-2.5
2.5-3
3-4
> 4
Speed Unknown
Lynn Canal
Gustavus
OLGA-NEVA STRAITS
Pelican
Hoonah
CHICHAGOF IS.
Juneau
Neva Strait
Tenakee
Olga Strait
PERIL STRAIT
Seymour Canal
ADMIRALTY IS.
BARANOF IS.
Angoon
Sitka
N

Shaded Areas: See SELECTED CHANNELS

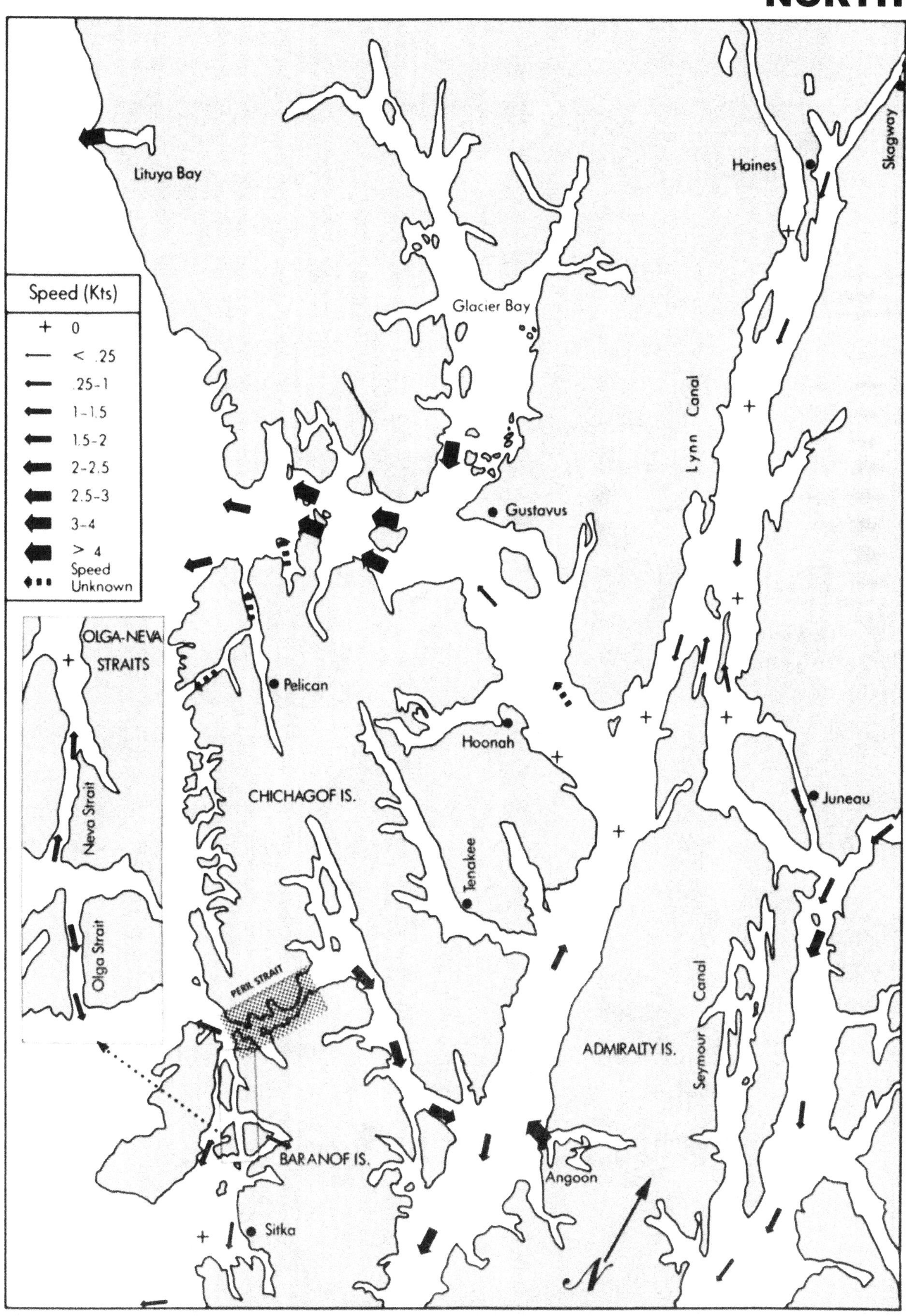
Lituya Bay
Glacier Bay
Haines
Skagway
Gustavus
Lynn Canal
Speed (Kts)
+ 0
< .25
.25-1
1-1.5
1.5-2
2-2.5
2.5-3
3-4
> 4
Speed Unknown
OLGA-NEVA STRAITS
Neva Strait
Olga Strait
Pelican
Hoonah
CHICHAGOF IS.
Juneau
Tenakee
PERIL STRAIT
Seymour Canal
ADMIRALTY IS.
BARANOF IS.
Angoon
Sitka
N

Shaded Areas: See SELECTED CHANNELS

NORTH

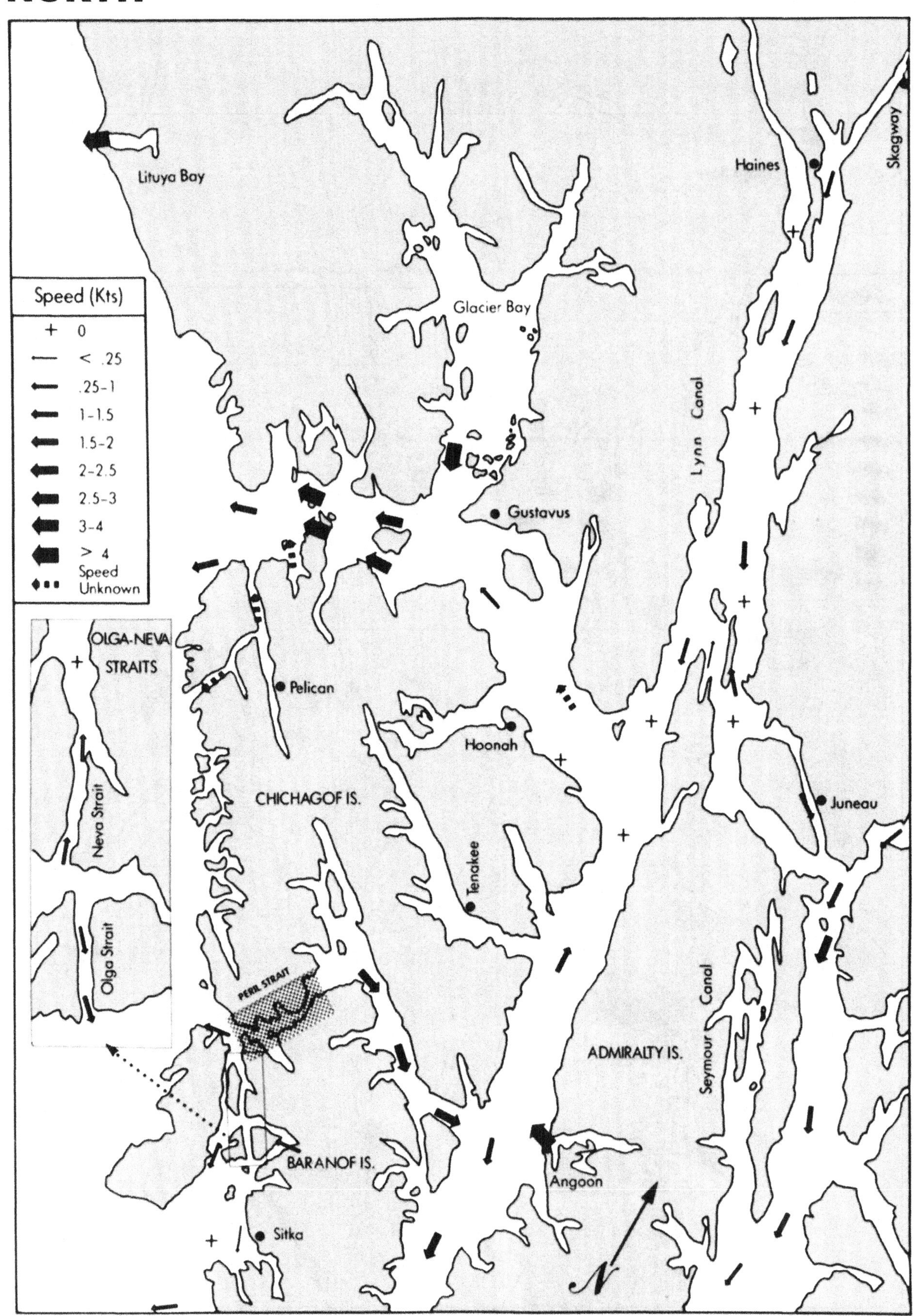

Lituya Bay
Haines
Skagway
Speed (Kts)
0
< .25
.25-1
1-1.5
1.5-2
2-2.5
2.5-3
3-4
> 4
Speed Unknown
Glacier Bay
Lynn Canal
Gustavus
OLGA-NEVA STRAITS
Pelican
Hoonah
CHICHAGOF IS.
Juneau
Neva Strait
Tenakee
Olga Strait
PERIL STRAIT
Seymour Canal
ADMIRALTY IS.
BARANOF IS.
Angoon
Sitka
N

Shaded Areas: See SELECTED CHANNELS

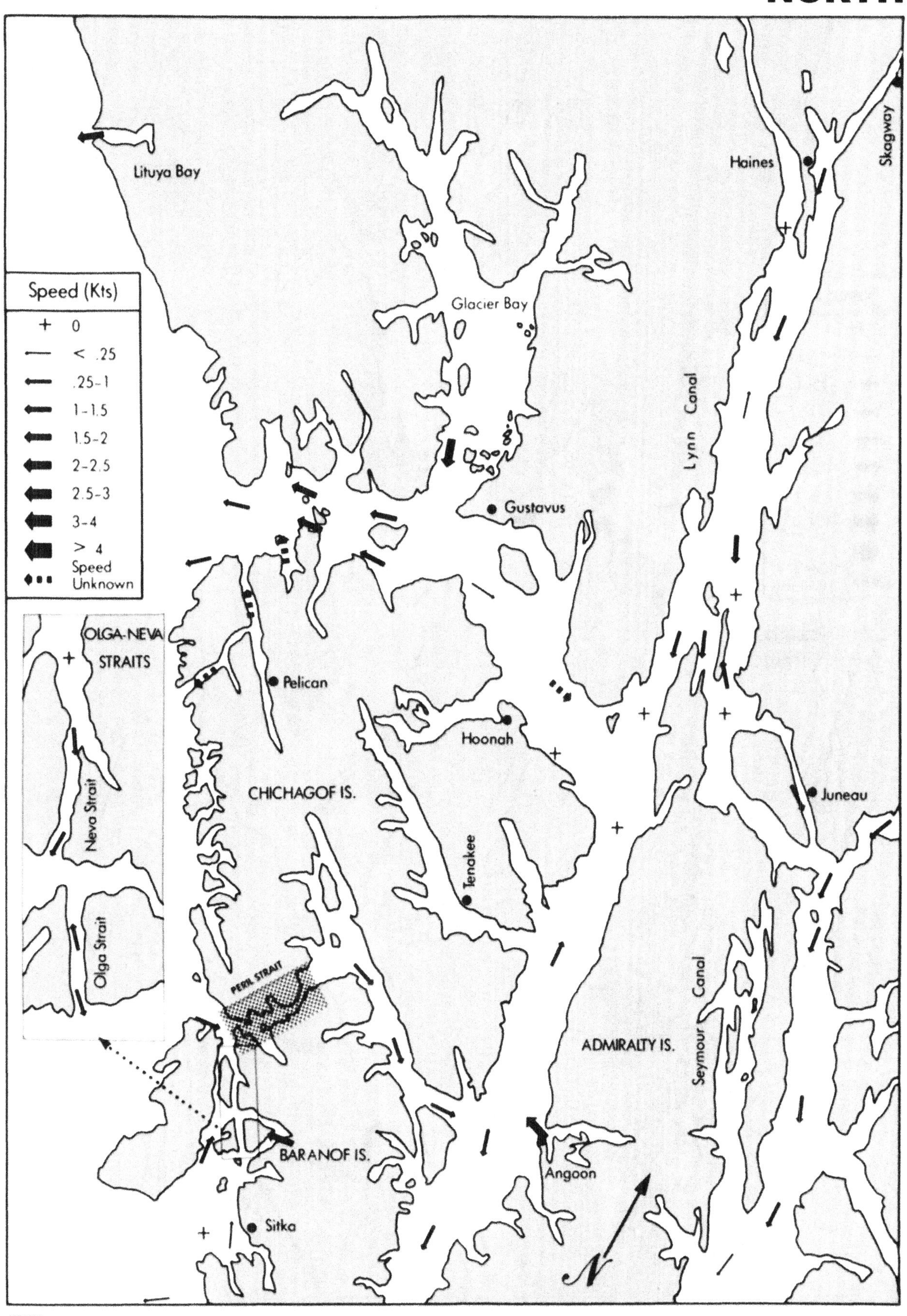

Lituya Bay
Haines
Skagway
Speed (Kts)
+ 0
< .25
.25-1
1-1.5
1.5-2
2-2.5
2.5-3
3-4
> 4
Speed Unknown
Glacier Bay
Lynn Canal
Gustavus
OLGA-NEVA STRAITS
Pelican
Hoonah
CHICHAGOF IS.
Juneau
Neva Strait
Tenakee
Olga Strait
PERIL STRAIT
Seymour Canal
ADMIRALTY IS.
BARANOF IS.
Angoon
Sitka
N

Shaded Areas: See SELECTED CHANNELS

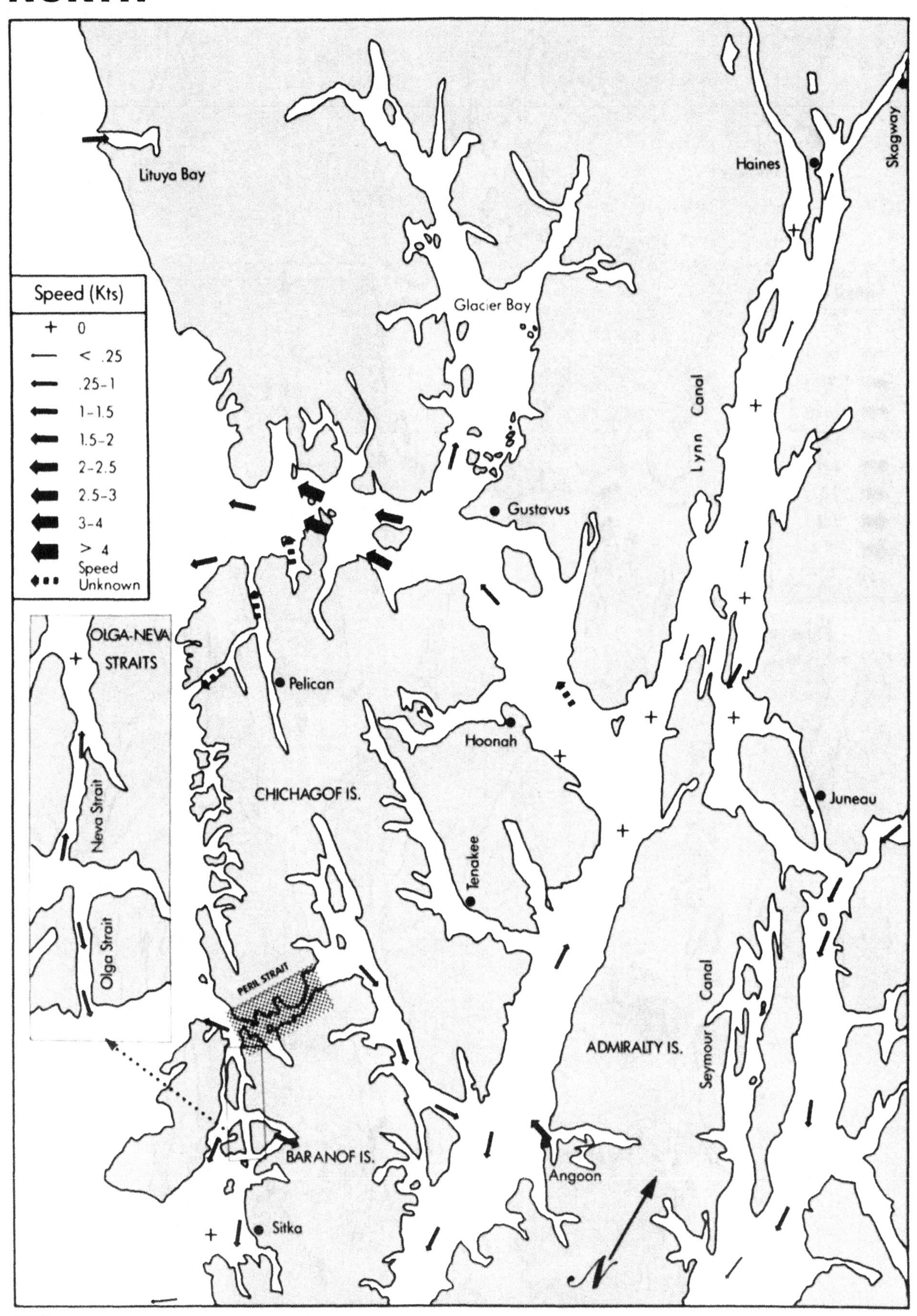
Lituya Bay
Glacier Bay
Haines
Skagway
Speed (Kts)
0
< .25
.25-1
1-1.5
1.5-2
2-2.5
2.5-3
3-4
> 4
Speed Unknown
Gustavus
Lynn Canal
OLGA-NEVA STRAITS
Pelican
Hoonah
Neva Strait
Olga Strait
CHICHAGOF IS.
Juneau
Tenakee
PERIL STRAIT
ADMIRALTY IS.
Seymour Canal
BARANOF IS.
Angoon
Sitka
N

Shaded Areas: See SELECTED CHANNELS

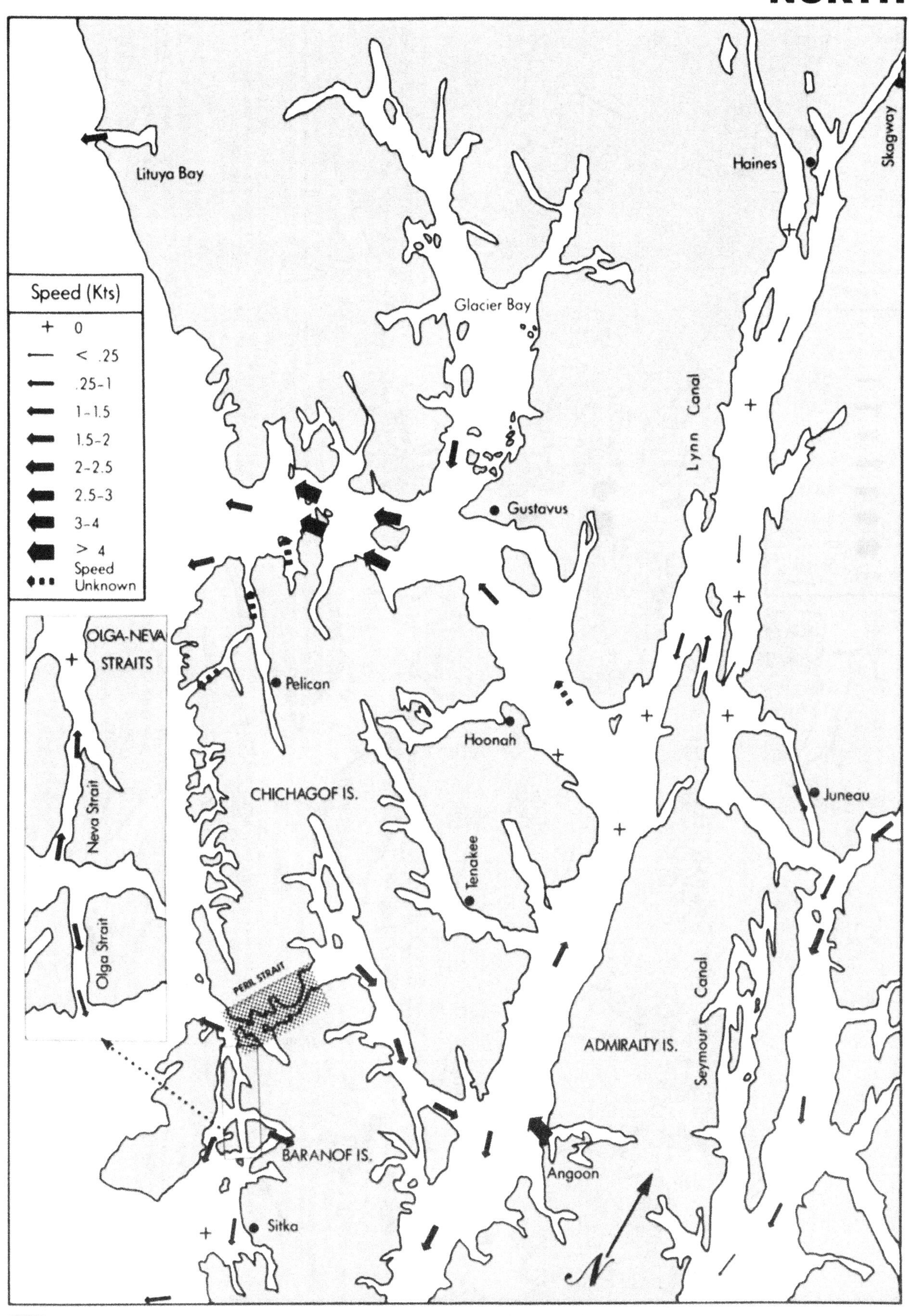
Lituya Bay
Haines
Skagway
Glacier Bay
Speed (Kts)
+ 0
< .25
.25-1
1-1.5
1.5-2
2-2.5
2.5-3
3-4
> 4
Speed Unknown
Gustavus
Lynn Canal
OLGA-NEVA STRAITS
Pelican
Hoonah
CHICHAGOF IS.
Juneau
Neva Strait
Tenakee
Olga Strait
PERIL STRAIT
Seymour Canal
ADMIRALTY IS.
BARANOF IS.
Angoon
Sitka
N

Shaded Areas: See SELECTED CHANNELS

NORTH

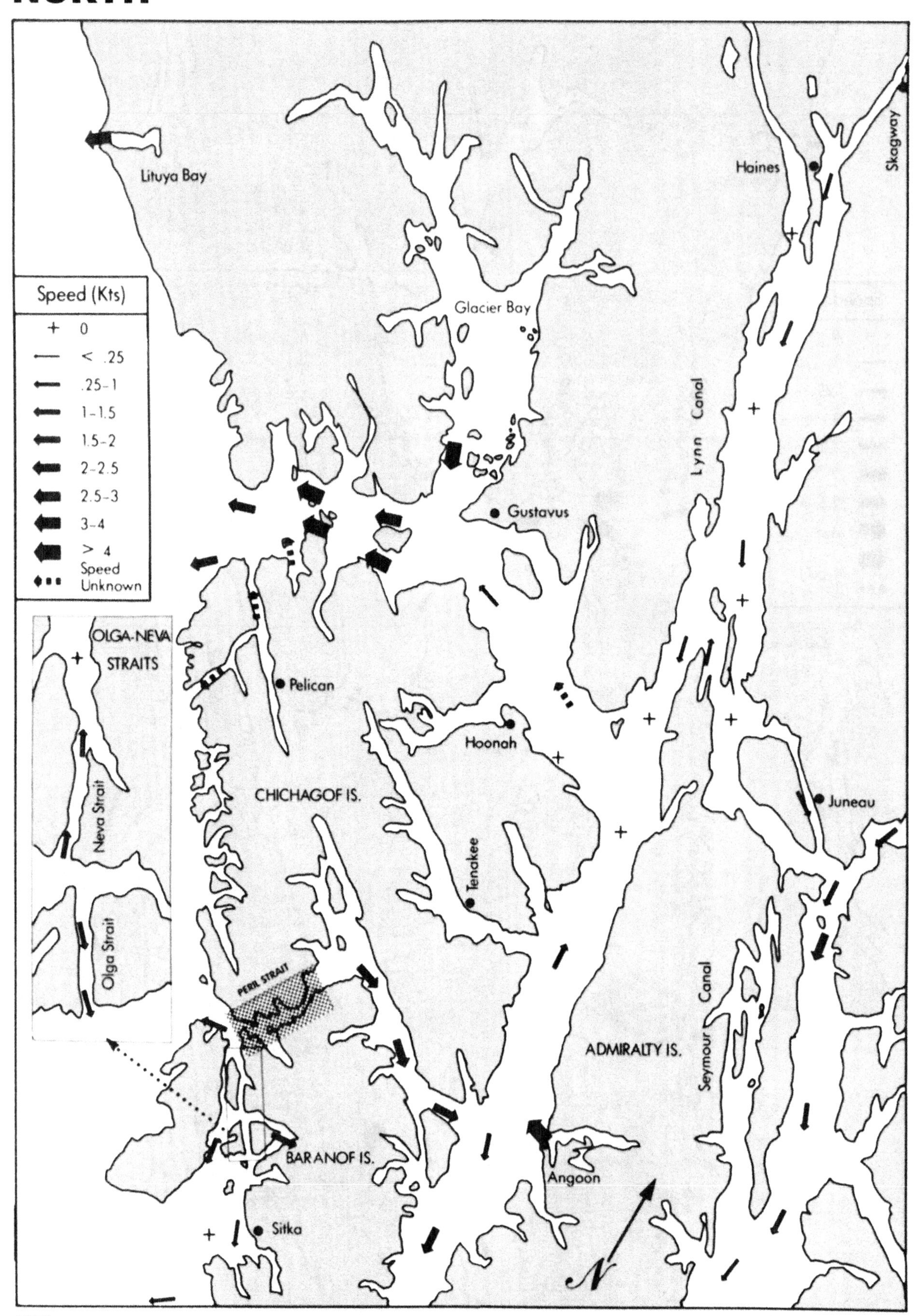

Speed (Kts)
+ 0
< .25
.25-1
1-1.5
1.5-2
2-2.5
2.5-3
3-4
> 4
Speed Unknown
Lituya Bay
Glacier Bay
Gustavus
Haines
Skagway
Lynn Canal
Pelican
Hoonah
Juneau
CHICHAGOF IS.
Tenakee
OLGA-NEVA STRAITS
Neva Strait
Olga Strait
PERIL STRAIT
ADMIRALTY IS.
Seymour Canal
BARANOF IS.
Angoon
Sitka
N

Shaded Areas: See SELECTED CHANNELS

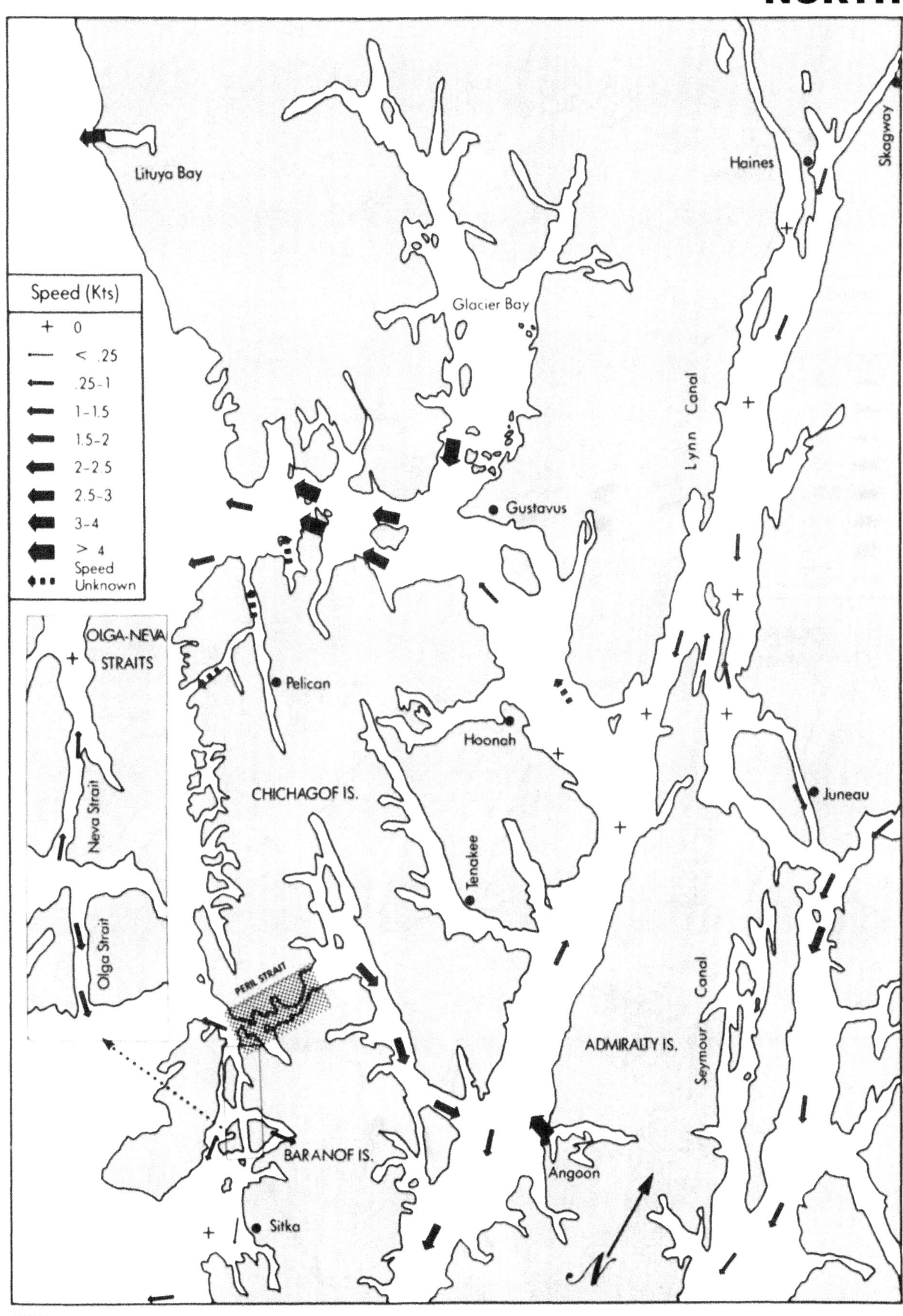
Lituya Bay
Haines
Skagway
Glacier Bay
Speed (Kts)
+ 0
< .25
.25-1
1-1.5
1.5-2
2-2.5
2.5-3
3-4
> 4
Speed Unknown
Lynn Canal
Gustavus
Pelican
Hoonah
OLGA-NEVA STRAITS
Neva Strait
Olga Strait
CHICHAGOF IS.
Juneau
Tenakee
PERIL STRAIT
ADMIRALTY IS.
Seymour Canal
BARANOF IS.
Angoon
Sitka

Shaded Areas: See SELECTED CHANNELS

NORTH

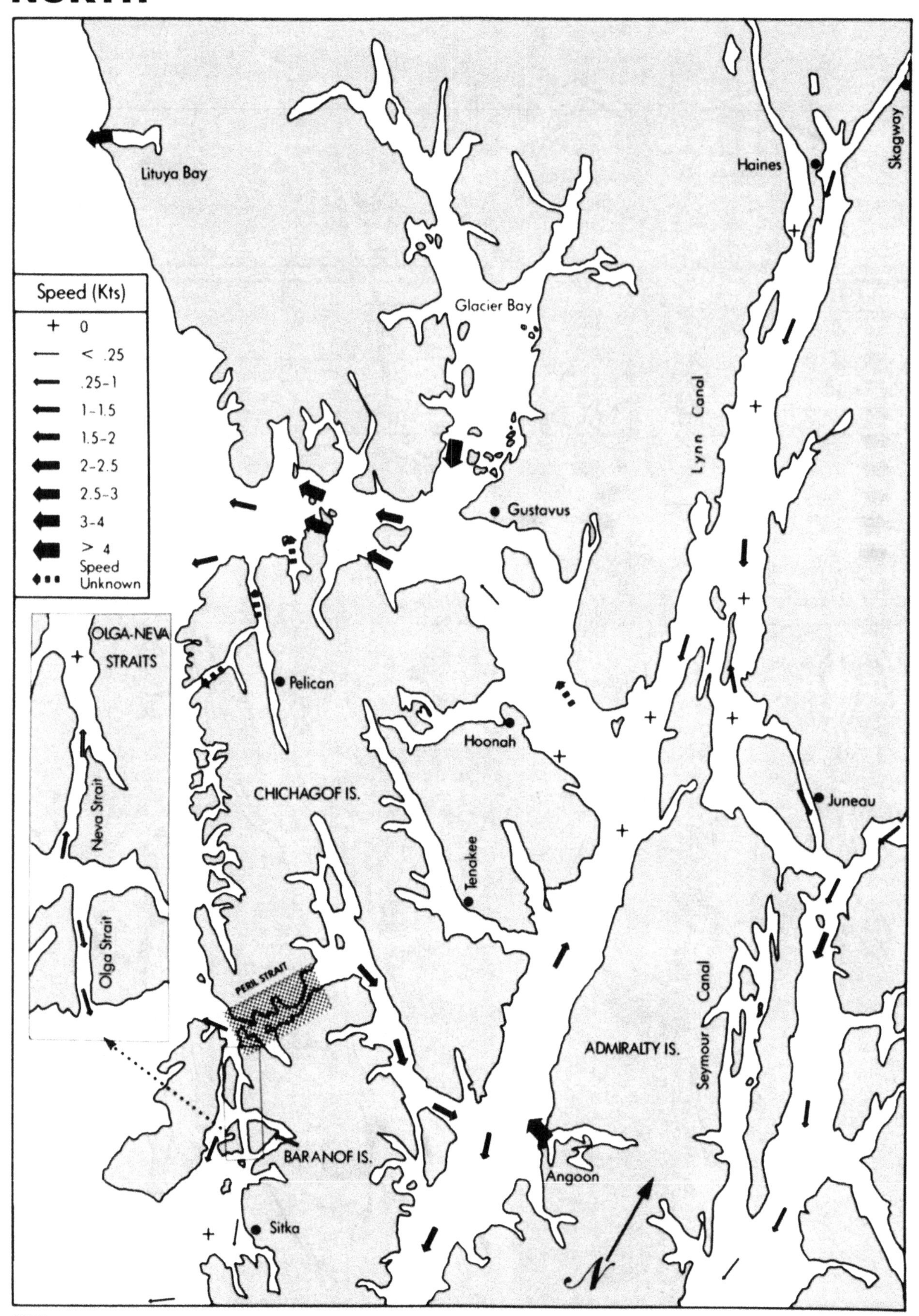

Speed (Kts)
+ 0
< .25
.25-1
1-1.5
1.5-2
2-2.5
2.5-3
3-4
> 4
Speed Unknown
Lituya Bay
Glacier Bay
Haines
Skagway
Lynn Canal
Gustavus
Pelican
Hoonah
CHICHAGOF IS.
Tenakee
Juneau
OLGA-NEVA STRAITS
Neva Strait
Olga Strait
PERIL STRAIT
ADMIRALTY IS.
Seymour Canal
BARANOF IS.
Angoon
Sitka

Shaded Areas: See SELECTED CHANNELS

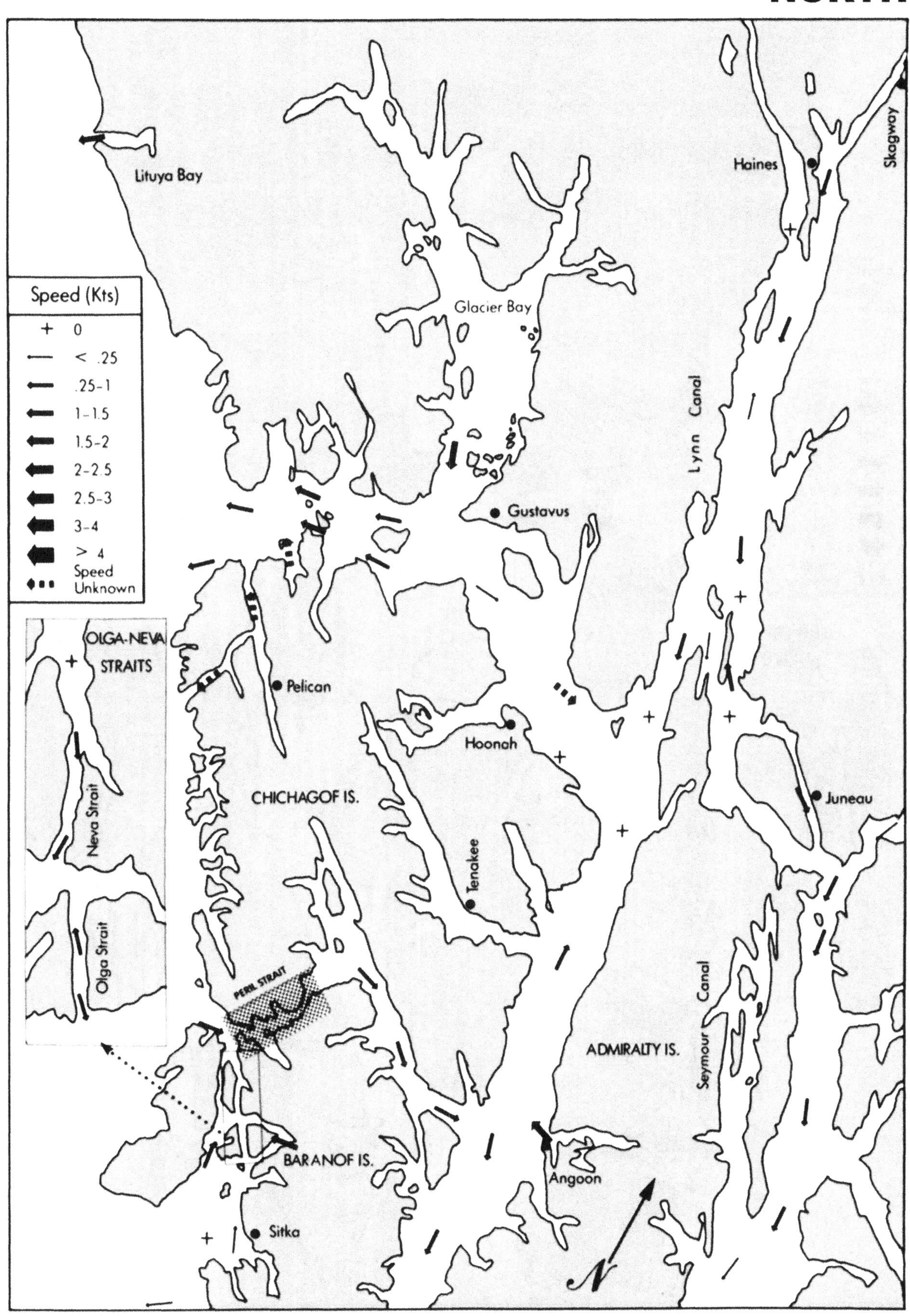
Speed (Kts)
+ 0
< .25
.25-1
1-1.5
1.5-2
2-2.5
2.5-3
3-4
> 4
Speed Unknown
Lituya Bay
Glacier Bay
Haines
Skagway
Lynn Canal
Gustavus
Pelican
Hoonah
Juneau
CHICHAGOF IS.
Tenakee
OLGA-NEVA STRAITS
Neva Strait
Olga Strait
PERIL STRAIT
ADMIRALTY IS.
Seymour Canal
BARANOF IS.
Angoon
Sitka
N

Shaded Areas: See SELECTED CHANNELS

NORTH

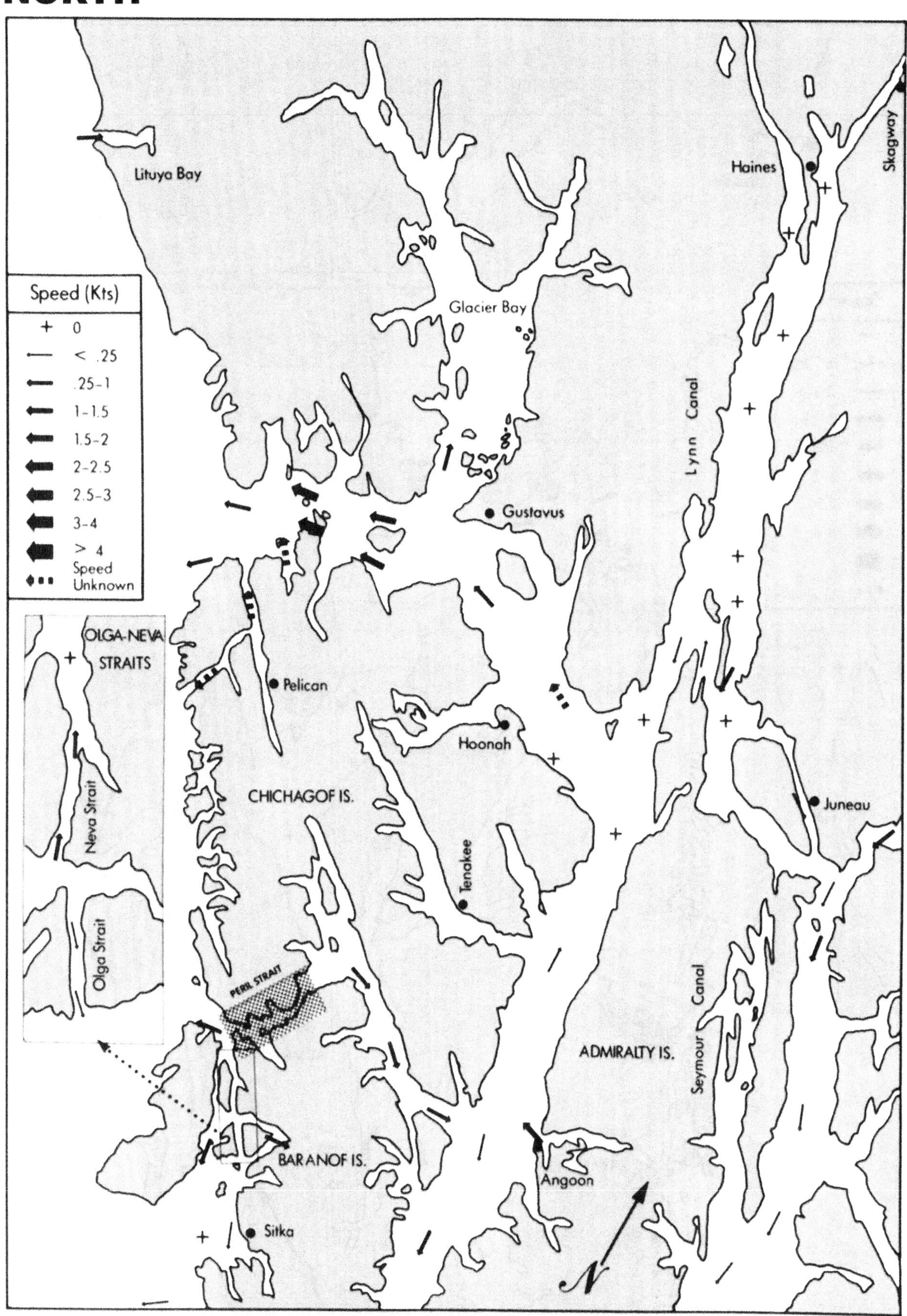
Lituya Bay
Glacier Bay
Haines
Skagway
Speed (Kts)
+ 0
< .25
.25-1
1-1.5
1.5-2
2-2.5
2.5-3
3-4
> 4
Speed Unknown
Lynn Canal
Gustavus
OLGA-NEVA STRAITS
Pelican
Hoonah
Neva Strait
CHICHAGOF IS.
Juneau
Tenakee
Olga Strait
PERIL STRAIT
Seymour Canal
ADMIRALTY IS.
BARANOF IS.
Angoon
Sitka

Shaded Areas: See SELECTED CHANNELS

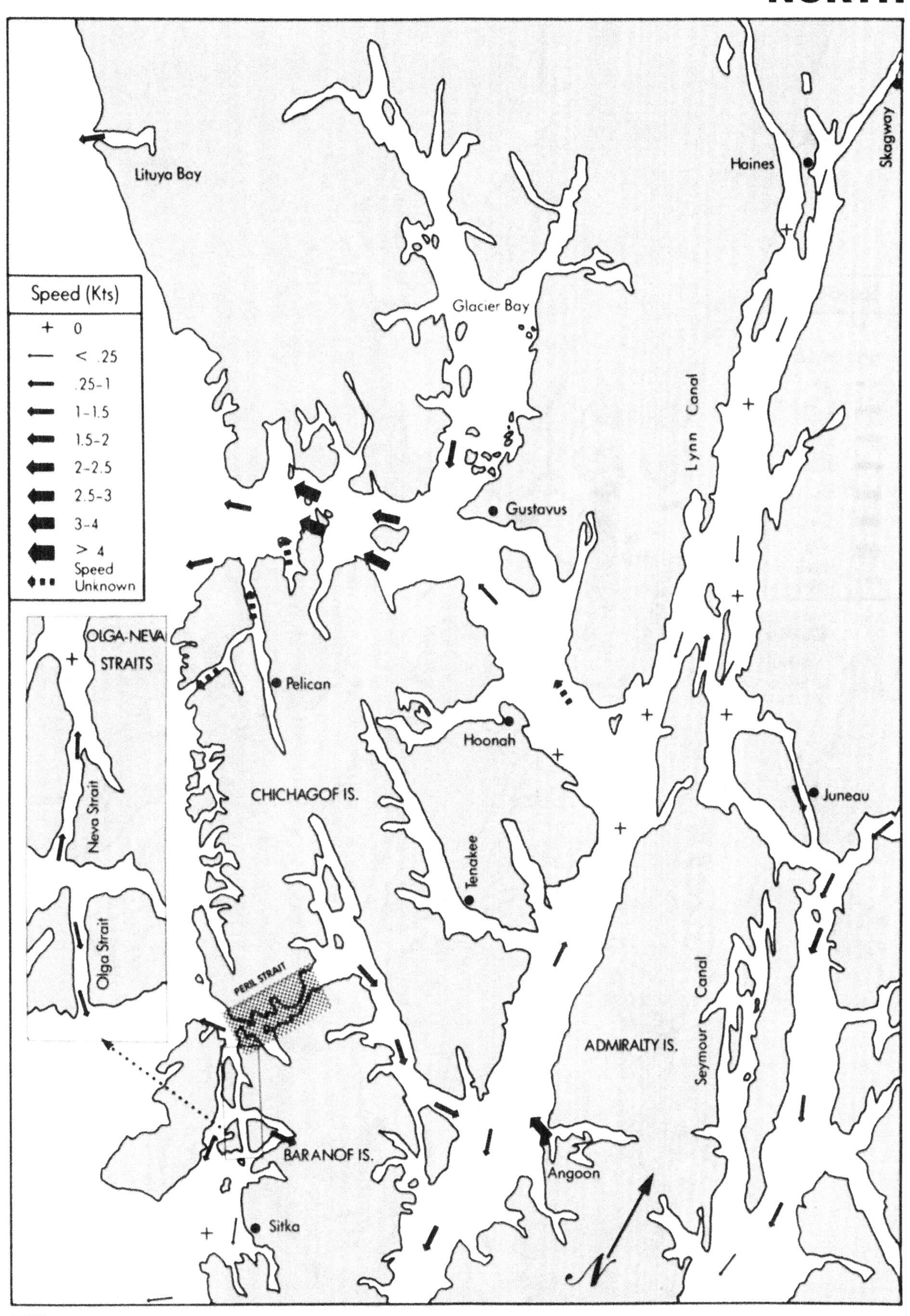
Lituya Bay
Glacier Bay
Haines
Skagway
Gustavus
Lynn Canal
Pelican
Hoonah
Juneau
CHICHAGOF IS.
Tenakee
PERIL STRAIT
ADMIRALTY IS.
Seymour Canal
BARANOF IS.
Angoon
Sitka
Speed (Kts)
+ 0
< .25
.25-1
1-1.5
1.5-2
2-2.5
2.5-3
3-4
> 4
Speed Unknown
OLGA-NEVA STRAITS
Neva Strait
Olga Strait
N

Shaded Areas: See SELECTED CHANNELS

NORTH

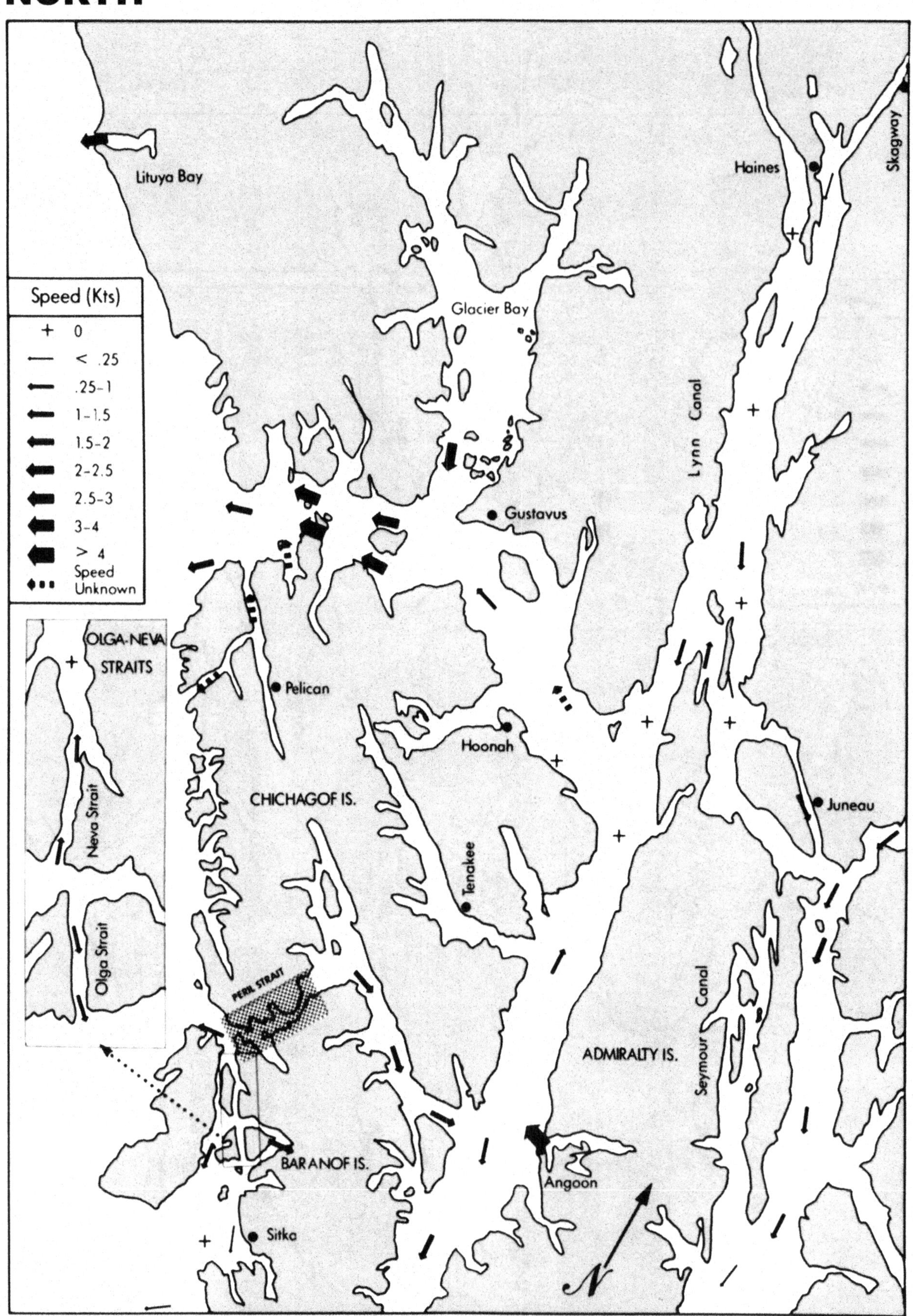

Lituya Bay
Glacier Bay
Haines
Skagway
Speed (Kts)
0
< .25
.25–1
1–1.5
1.5–2
2–2.5
2.5–3
3–4
> 4
Speed Unknown
Gustavus
Lynn Canal
OLGA-NEVA STRAITS
Pelican
Hoonah
Juneau
Neva Strait
CHICHAGOF IS.
Tenakee
Olga Strait
PERIL STRAIT
Seymour Canal
ADMIRALTY IS.
BARANOF IS.
Angoon
Sitka
N

Shaded Areas: See SELECTED CHANNELS

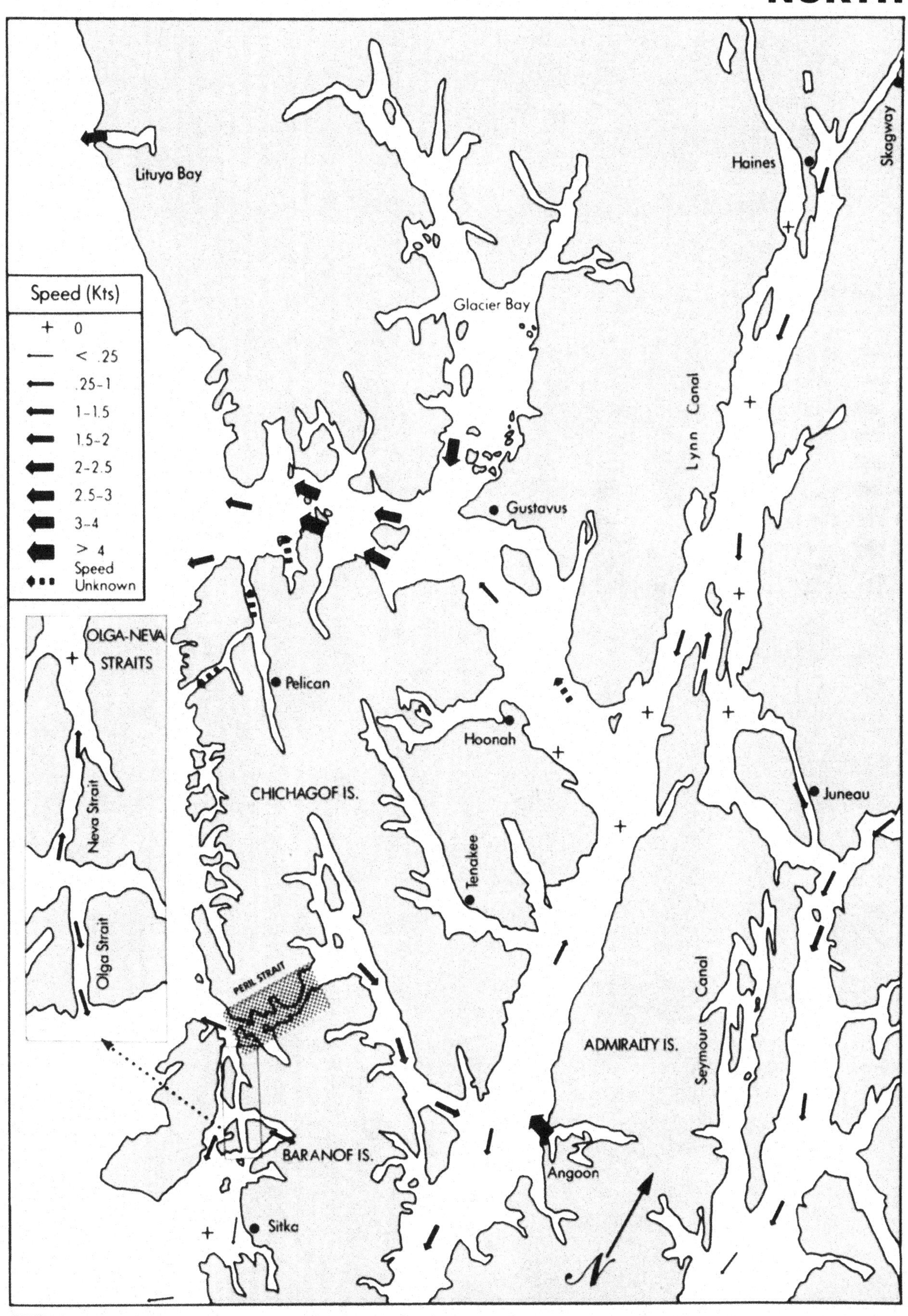

Lituya Bay
Speed (Kts)
0
< .25
.25-1
1-1.5
1.5-2
2-2.5
2.5-3
3-4
> 4
Speed Unknown
Glacier Bay
Gustavus
Haines
Skagway
Lynn Canal
Pelican
Hoonah
Juneau
CHICHAGOF IS.
Tenakee
OLGA-NEVA STRAITS
Neva Strait
Olga Strait
PERIL STRAIT
ADMIRALTY IS.
Seymour Canal
BARANOF IS.
Angoon
Sitka

Shaded Areas: See SELECTED CHANNELS

NORTH

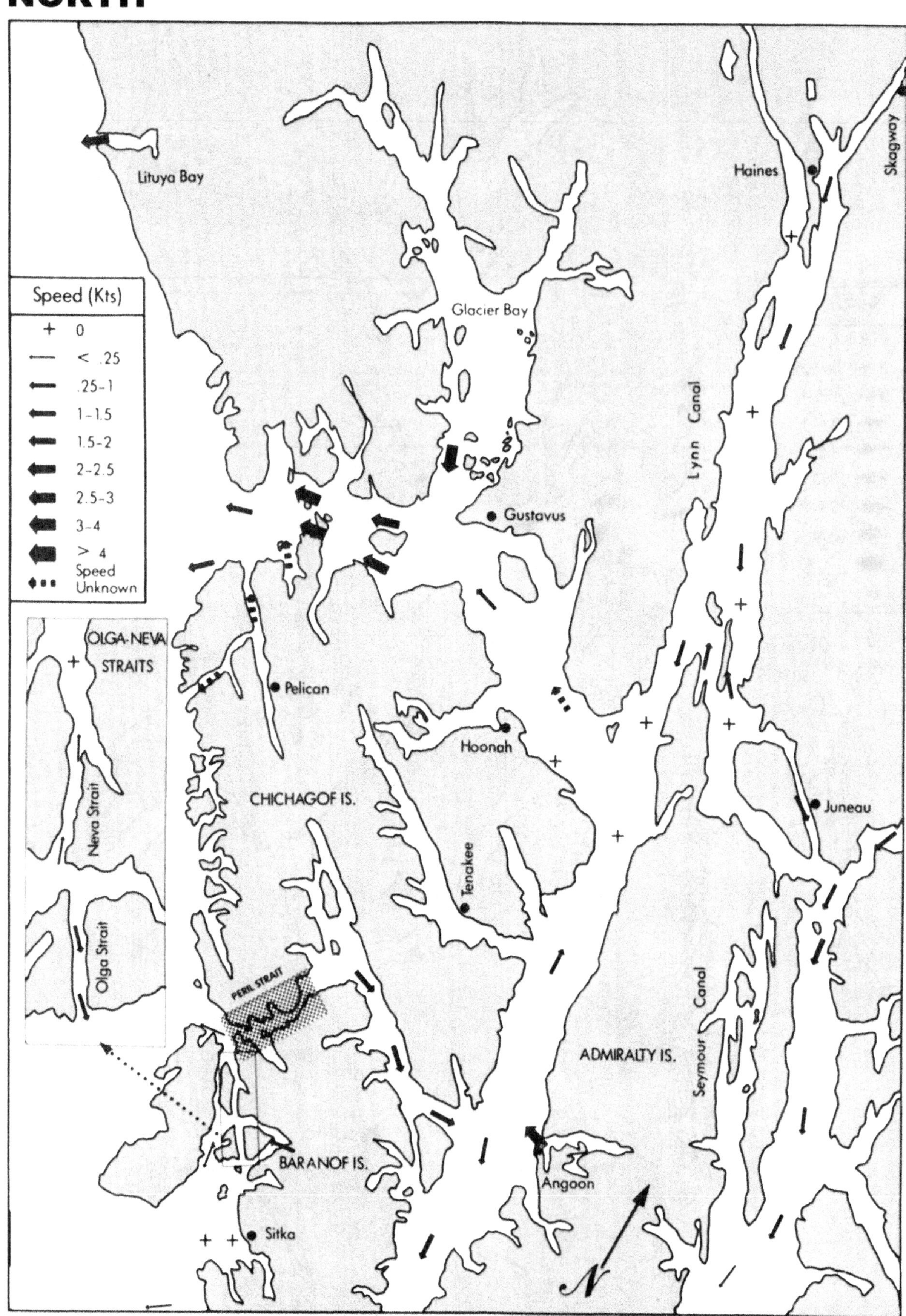

Speed (Kts)
0
< .25
.25-1
1-1.5
1.5-2
2-2.5
2.5-3
3-4
> 4
Speed Unknown
Lituya Bay
Glacier Bay
Haines
Skagway
Lynn Canal
Gustavus
Pelican
Hoonah
Juneau
CHICHAGOF IS.
Tenakee
OLGA-NEVA STRAITS
Neva Strait
Olga Strait
PERIL STRAIT
ADMIRALTY IS.
Seymour Canal
BARANOF IS.
Angoon
Sitka

Shaded Areas: See SELECTED CHANNELS

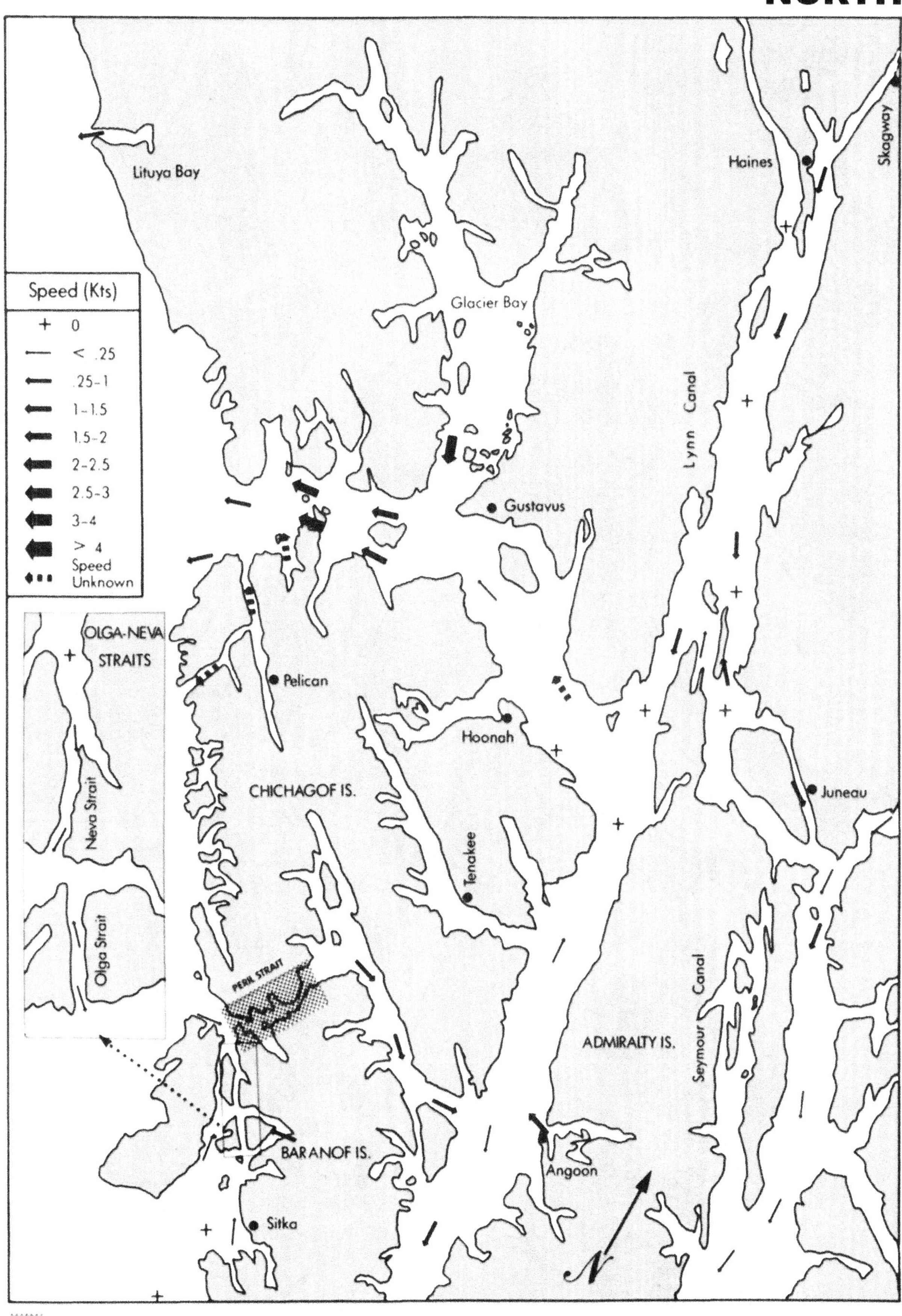
Lituya Bay
Haines
Skagway
Glacier Bay
Speed (Kts)
0
< .25
.25-1
1-1.5
1.5-2
2-2.5
2.5-3
3-4
> 4
Speed Unknown
Gustavus
Lynn Canal
OLGA-NEVA STRAITS
Pelican
Hoonah
Juneau
Neva Strait
Olga Strait
CHICHAGOF IS.
Tenakee
PERIL STRAIT
Seymour Canal
ADMIRALTY IS.
BARANOF IS.
Angoon
Sitka
N

Shaded Areas: See SELECTED CHANNELS

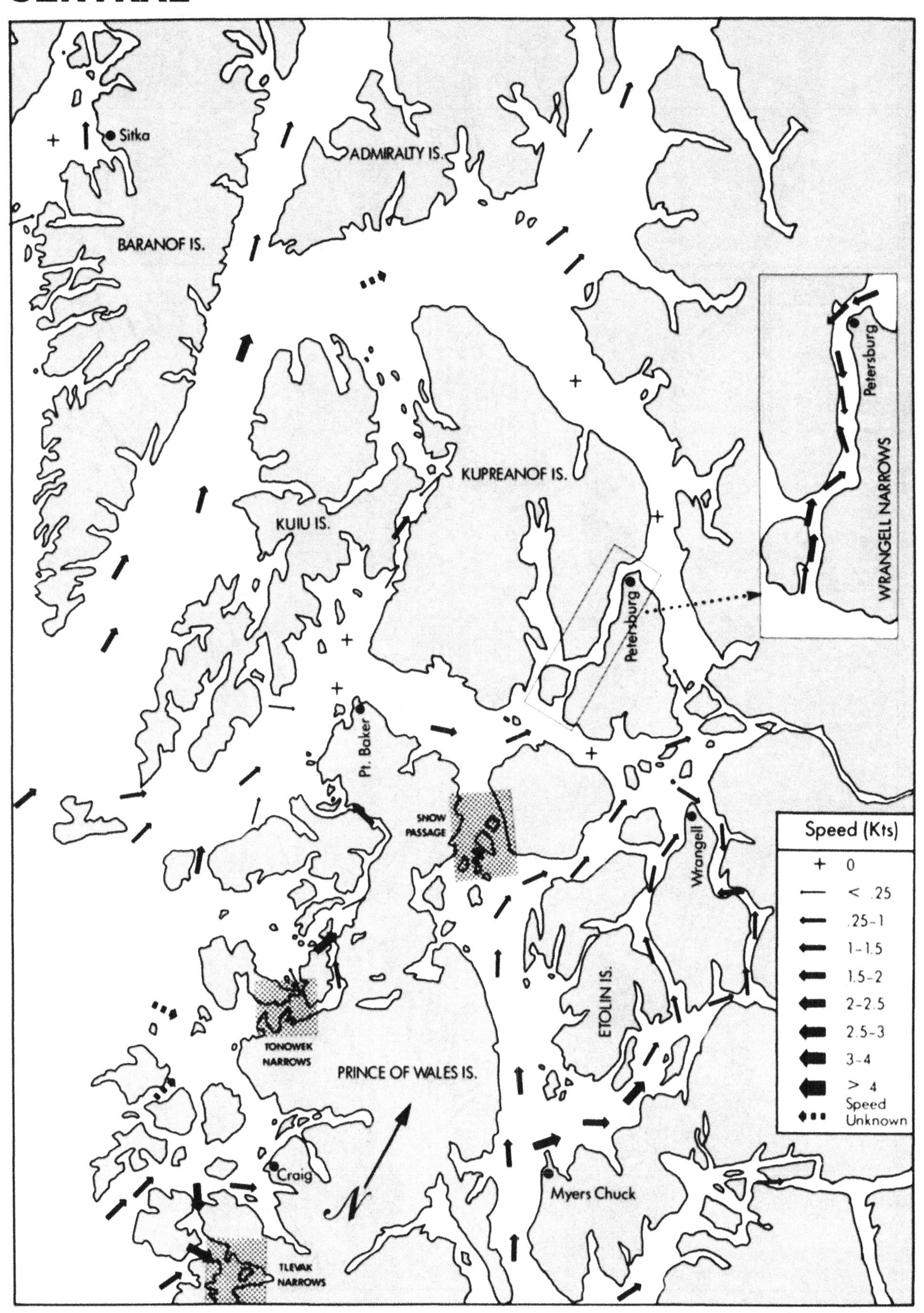
Sitka
ADMIRALTY IS.
BARANOF IS.
KUPREANOF IS.
KUIU IS.
Petersburg
Pt. Baker
SNOW PASSAGE
Wrangell
ETOLIN IS.
TONOWEK NARROWS
PRINCE OF WALES IS.
Craig
Myers Chuck
TLEVAK NARROWS
WRANGELL NARROWS
Speed (Kts)
0
< .25
.25–1
1–1.5
1.5–2
2–2.5
2.5–3
3–4
> 4
Speed Unknown

Shaded Areas: See SELECTED CHANNELS

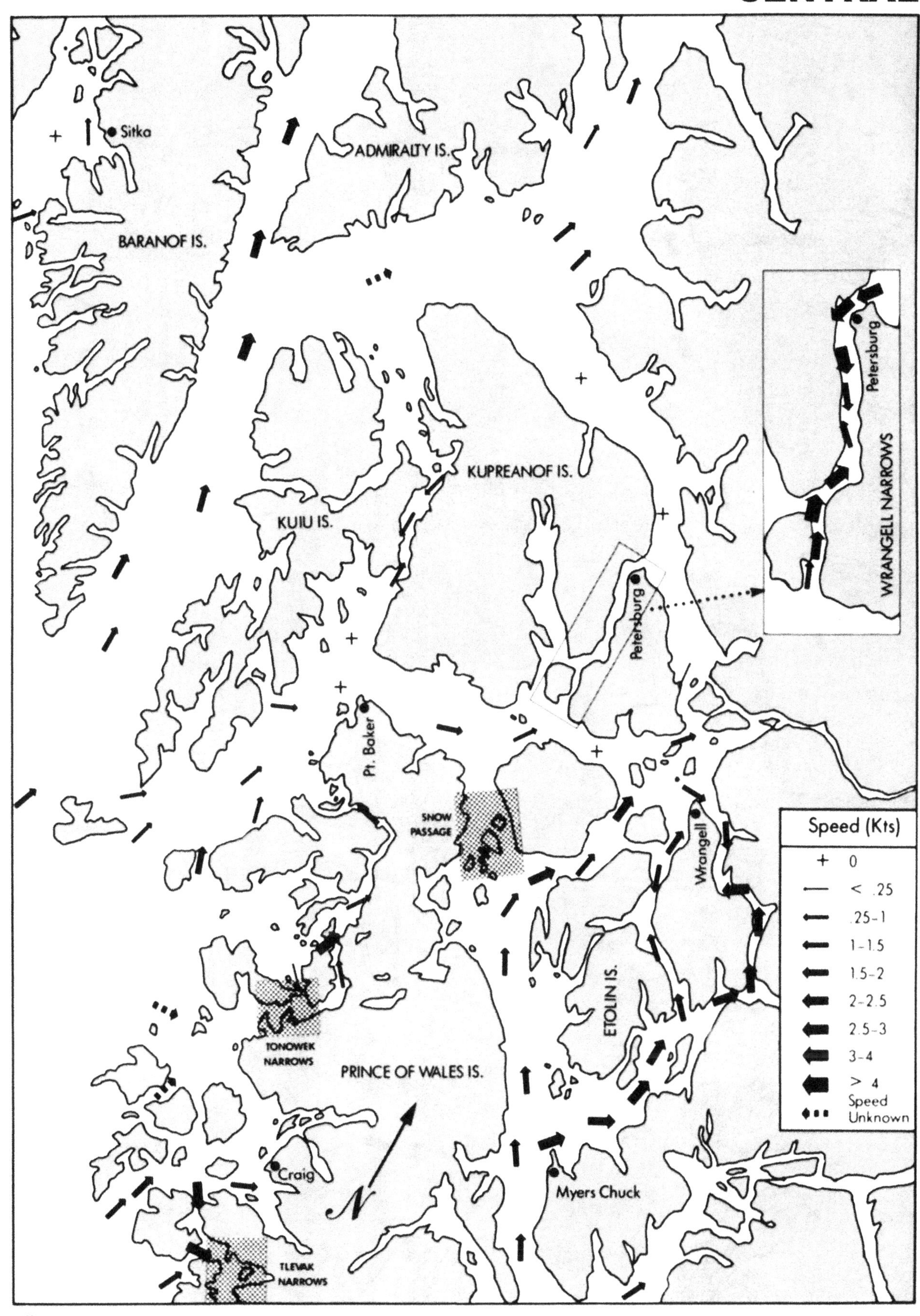

Sitka
ADMIRALTY IS.
BARANOF IS.
KUPREANOF IS.
KUIU IS.
Petersburg
WRANGELL NARROWS
Pt. Baker
SNOW PASSAGE
Wrangell
ETOLIN IS.
TONOWEK NARROWS
PRINCE OF WALES IS.
Craig
Myers Chuck
TLEVAK NARROWS
Speed (Kts)
+ 0
< .25
.25-1
1-1.5
1.5-2
2-2.5
2.5-3
3-4
> 4
Speed Unknown

Shaded Areas: See SELECTED CHANNELS

CENTRAL

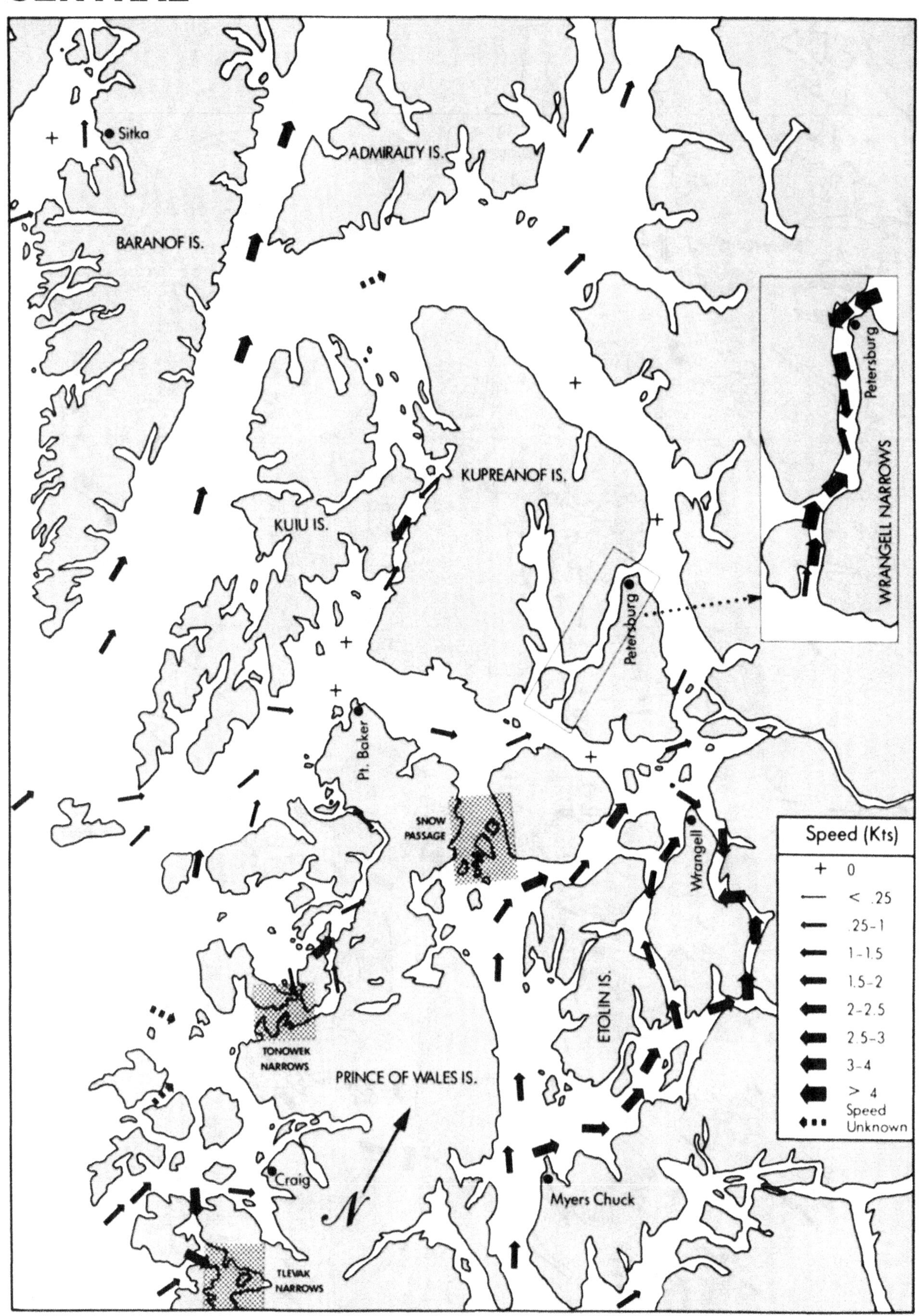

Sitka
ADMIRALTY IS.
BARANOF IS.
KUPREANOF IS.
KUIU IS.
Petersburg
Pt. Baker
SNOW PASSAGE
Wrangell
ETOLIN IS.
TONOWEK NARROWS
PRINCE OF WALES IS.
Craig
Myers Chuck
TLEVAK NARROWS
N
WRANGELL NARROWS
Petersburg
Speed (Kts)
0
< .25
.25-1
1-1.5
1.5-2
2-2.5
2.5-3
3-4
> 4
Speed Unknown

Shaded Areas: See SELECTED CHANNELS

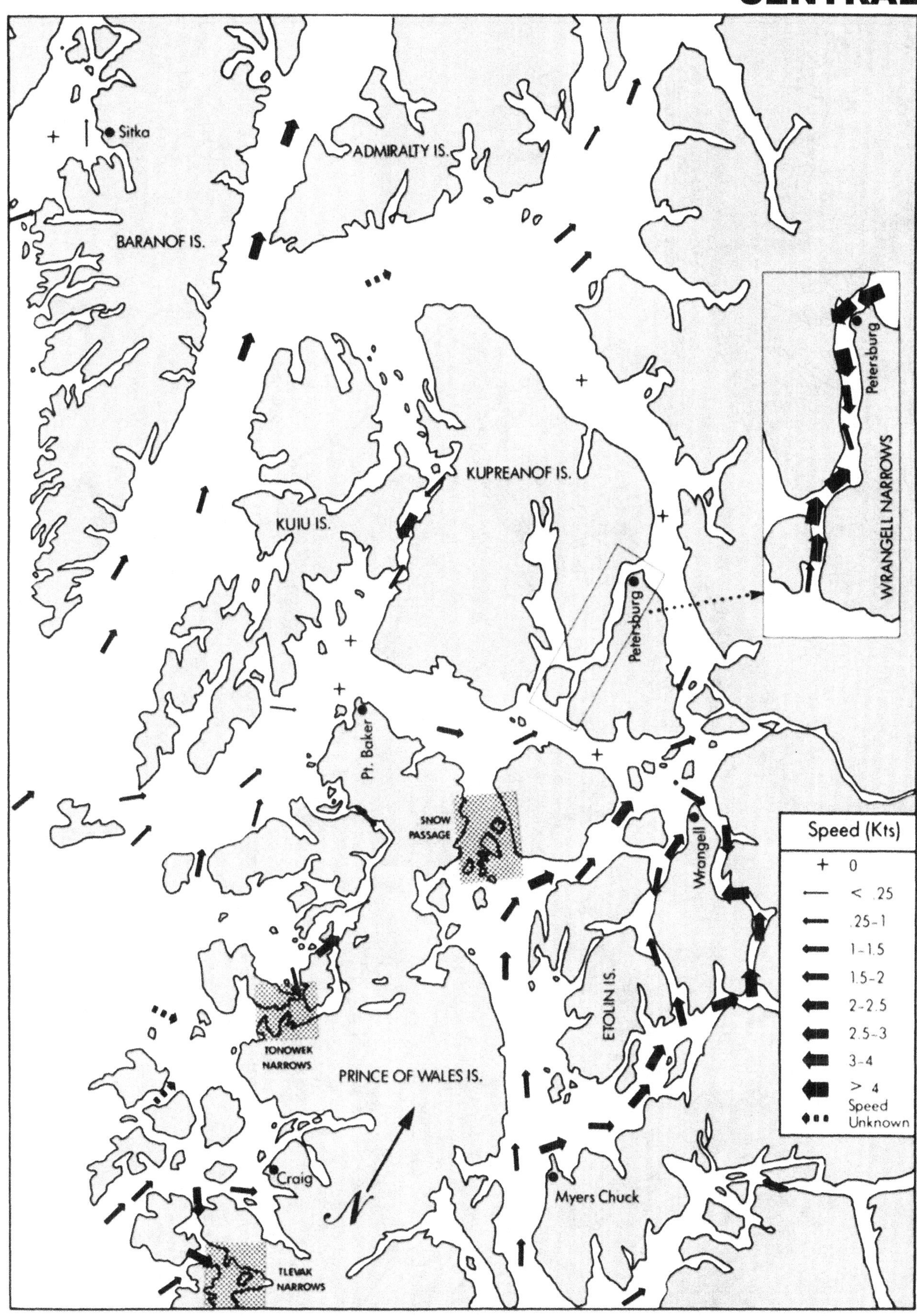
Sitka
ADMIRALTY IS.
BARANOF IS.
KUPREANOF IS.
KUIU IS.
Petersburg
WRANGELL NARROWS
Pt. Baker
SNOW PASSAGE
Wrangell
ETOLIN IS.
TONOWEK NARROWS
PRINCE OF WALES IS.
Craig
Myers Chuck
TLEVAK NARROWS
Speed (Kts)
0
< .25
.25-1
1-1.5
1.5-2
2-2.5
2.5-3
3-4
> 4
Speed Unknown

Shaded Areas: See SELECTED CHANNELS

CENTRAL

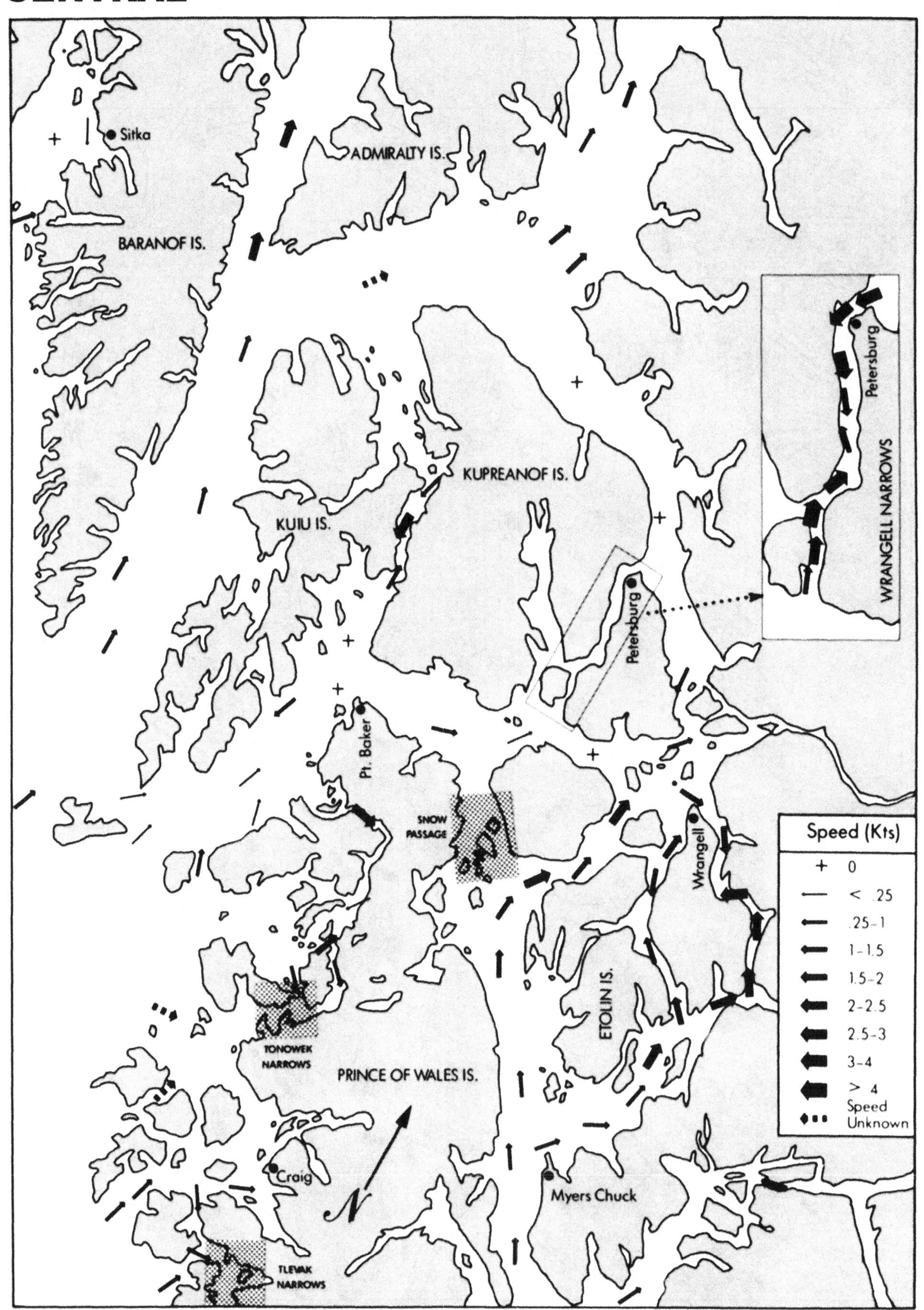

Sitka
ADMIRALTY IS.
BARANOF IS.
KUPREANOF IS.
KUIU IS.
Petersburg
WRANGELL NARROWS
Pt. Baker
SNOW PASSAGE
Wrangell
ETOLIN IS.
TONOWEK NARROWS
PRINCE OF WALES IS.
Craig
Myers Chuck
TLEVAK NARROWS
N
Speed (Kts)
0
< .25
.25-1
1-1.5
1.5-2
2-2.5
2.5-3
3-4
> 4
Speed Unknown

Shaded Areas: See SELECTED CHANNELS

5

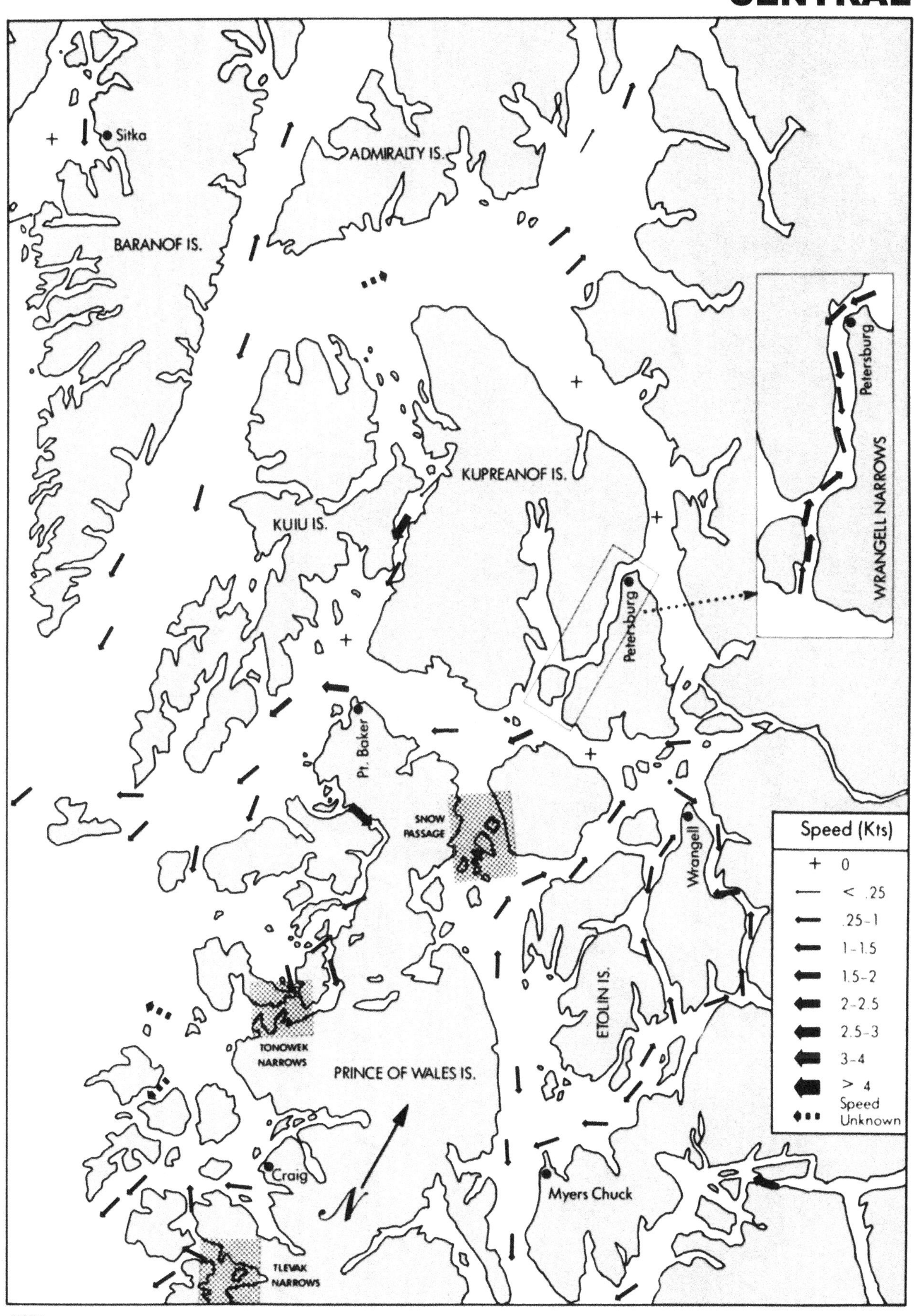
Sitka
ADMIRALTY IS.
BARANOF IS.
KUPREANOF IS.
KUIU IS.
Petersburg
WRANGELL NARROWS
Pt. Baker
SNOW PASSAGE
Wrangell
Speed (Kts)
0
< .25
.25-1
1-1.5
1.5-2
2-2.5
2.5-3
3-4
> 4
Speed Unknown
ETOLIN IS.
TONOWEK NARROWS
PRINCE OF WALES IS.
Craig
Myers Chuck
N
TLEVAK NARROWS

Shaded Areas: See SELECTED CHANNELS

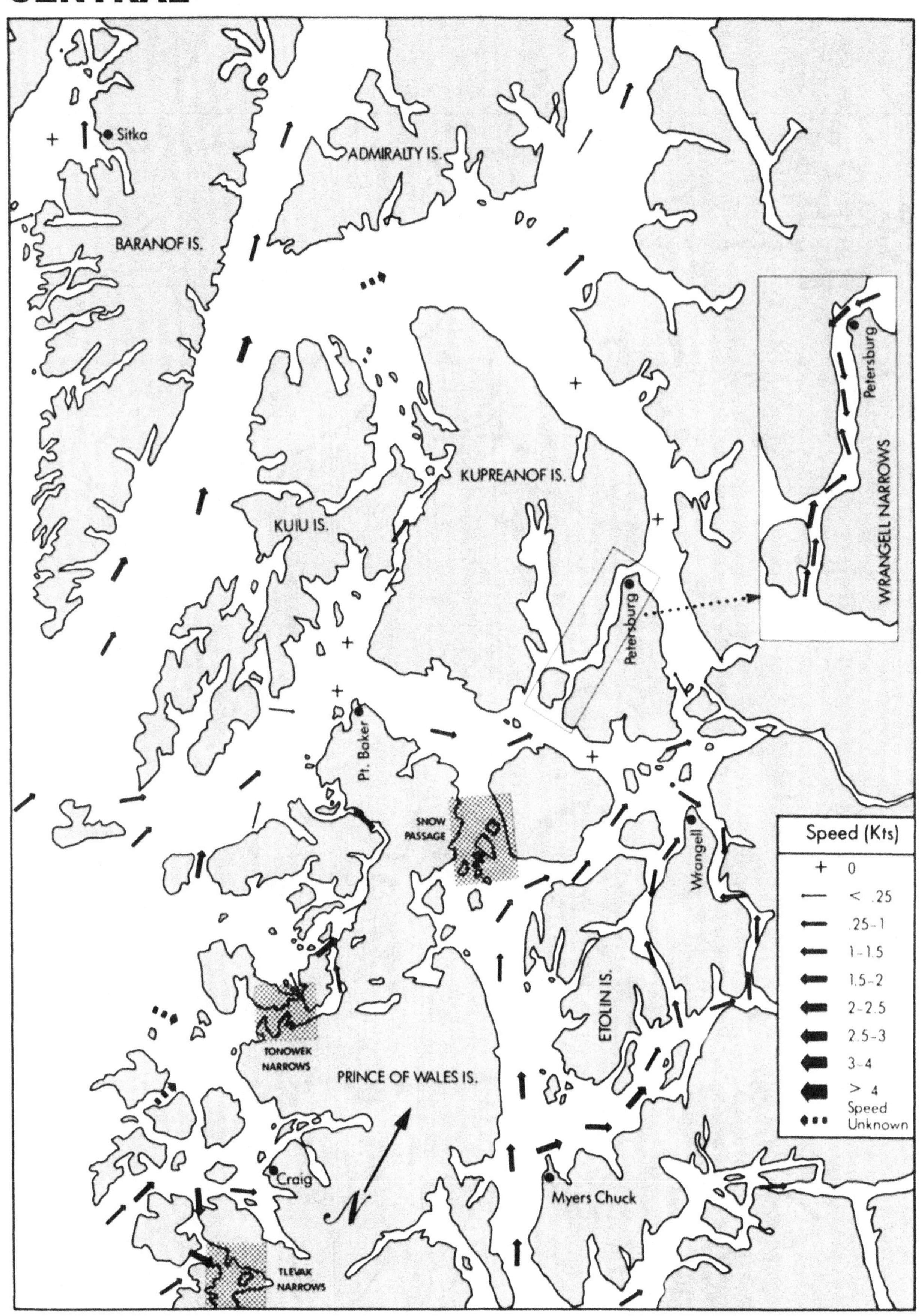

Sitka
ADMIRALTY IS.
BARANOF IS.
KUPREANOF IS.
KUIU IS.
Petersburg
WRANGELL NARROWS
Pt. Baker
SNOW PASSAGE
Wrangell
ETOLIN IS.
TONOWEK NARROWS
PRINCE OF WALES IS.
Craig
Myers Chuck
TLEVAK NARROWS
Speed (Kts)
0
< .25
.25-1
1-1.5
1.5-2
2-2.5
2.5-3
3-4
> 4
Speed Unknown

Shaded Areas: See SELECTED CHANNELS

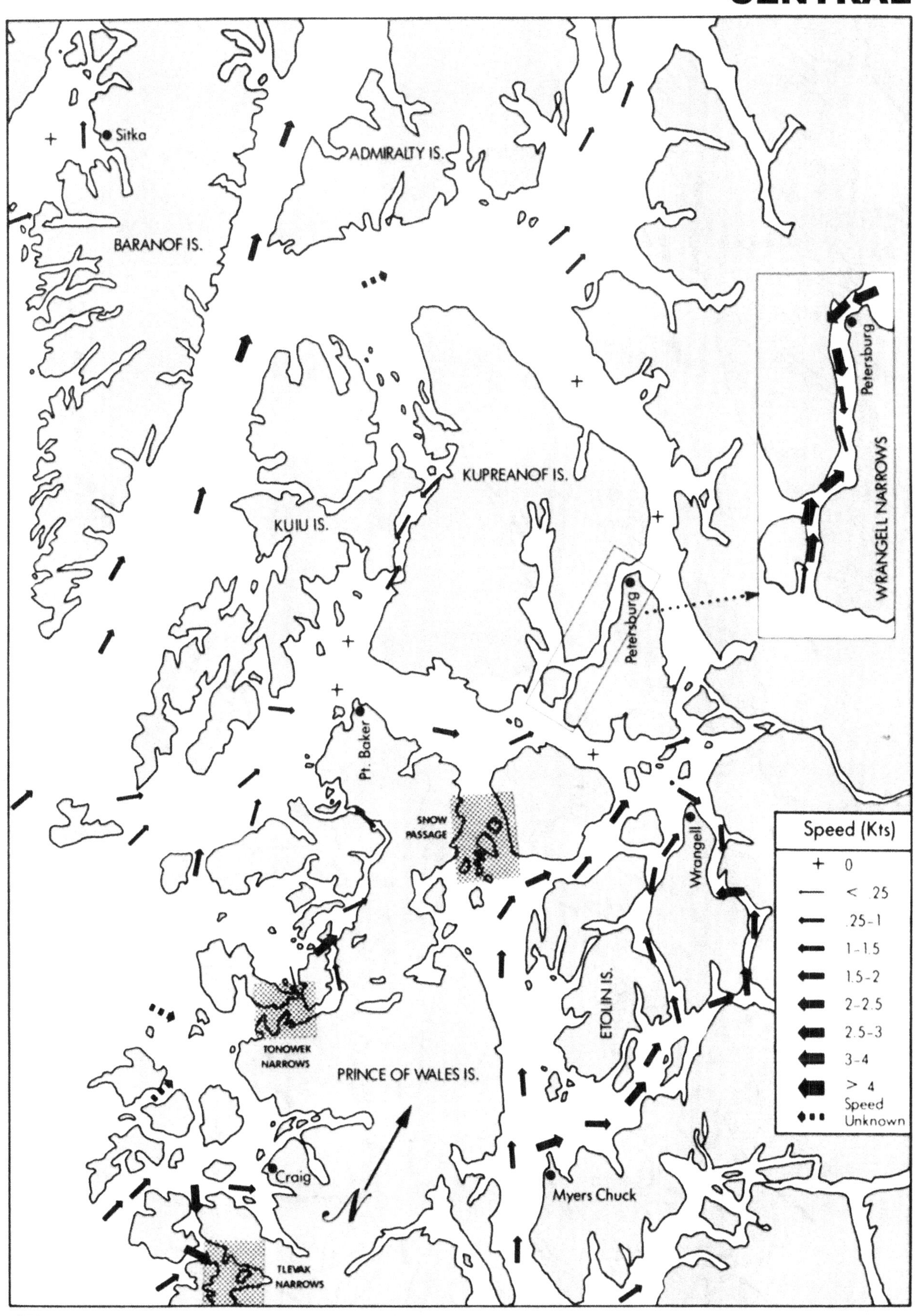
Sitka
ADMIRALTY IS.
BARANOF IS.
KUPREANOF IS.
KUIU IS.
Petersburg
Petersburg
WRANGELL NARROWS
Pt. Baker
SNOW PASSAGE
Wrangell
ETOLIN IS.
TONOWEK NARROWS
PRINCE OF WALES IS.
Craig
Myers Chuck
TLEVAK NARROWS
Speed (Kts)
0
< .25
.25-1
1-1.5
1.5-2
2-2.5
2.5-3
3-4
> 4
Speed Unknown

Shaded Areas: See SELECTED CHANNELS

CENTRAL

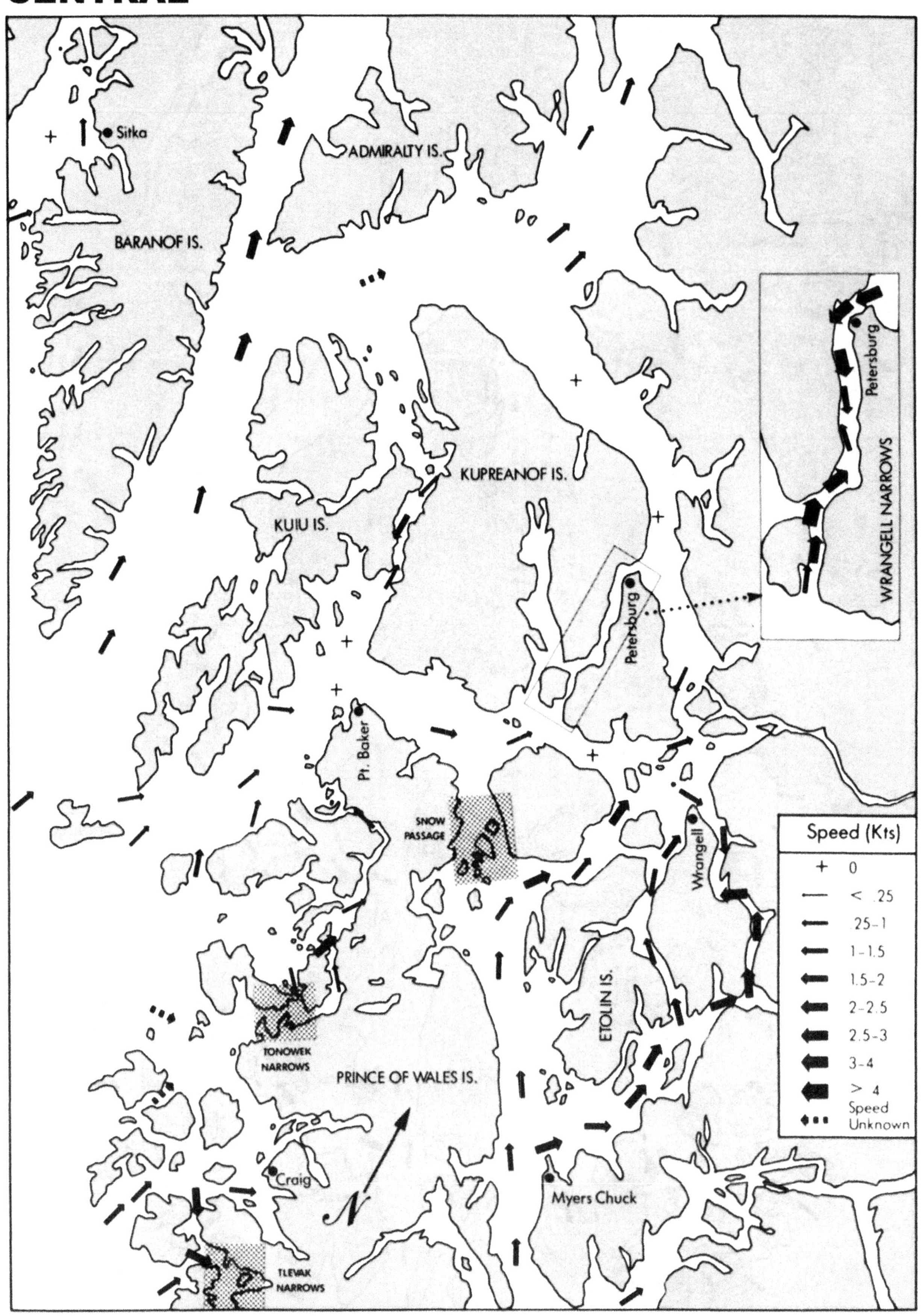

Sitka
ADMIRALTY IS.
BARANOF IS.
KUPREANOF IS.
KUIU IS.
Petersburg
WRANGELL NARROWS
Pt. Baker
SNOW PASSAGE
Wrangell
ETOLIN IS.
TONOWEK NARROWS
PRINCE OF WALES IS.
Craig
Myers Chuck
TLEVAK NARROWS
N
Speed (Kts)
+ 0
< .25
.25-1
1-1.5
1.5-2
2-2.5
2.5-3
3-4
> 4
Speed Unknown

Shaded Areas: See SELECTED CHANNELS

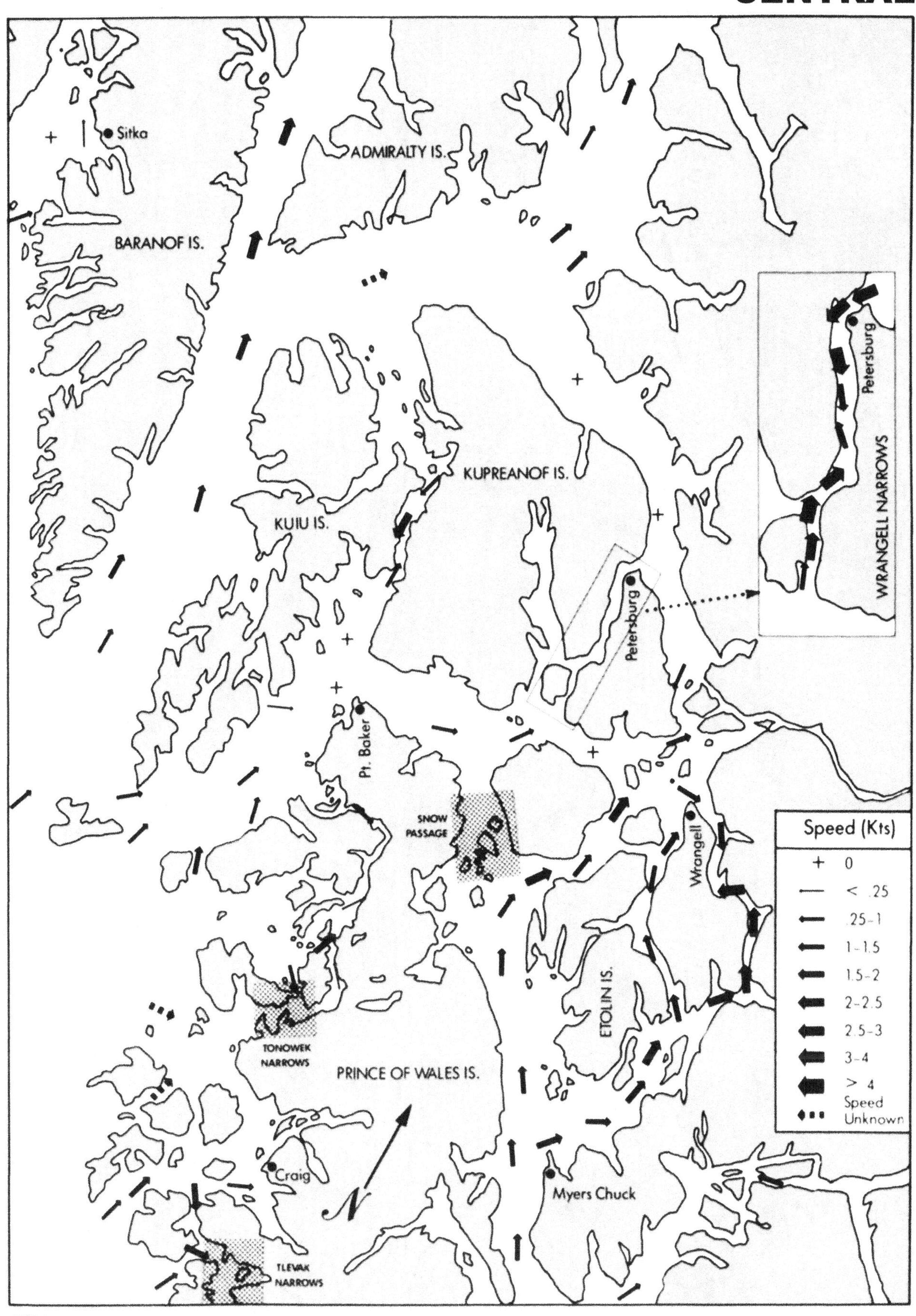
Sitka
ADMIRALTY IS.
BARANOF IS.
KUPREANOF IS.
KUIU IS.
Petersburg
WRANGELL NARROWS
Petersburg
Pt. Baker
SNOW PASSAGE
Wrangell
Speed (Kts)
+ 0
< .25
.25-1
1-1.5
1.5-2
2-2.5
2.5-3
3-4
> 4
Speed Unknown
ETOLIN IS.
TONOWEK NARROWS
PRINCE OF WALES IS.
Craig
N
Myers Chuck
TLEVAK NARROWS

Shaded Areas: See SELECTED CHANNELS

CENTRAL

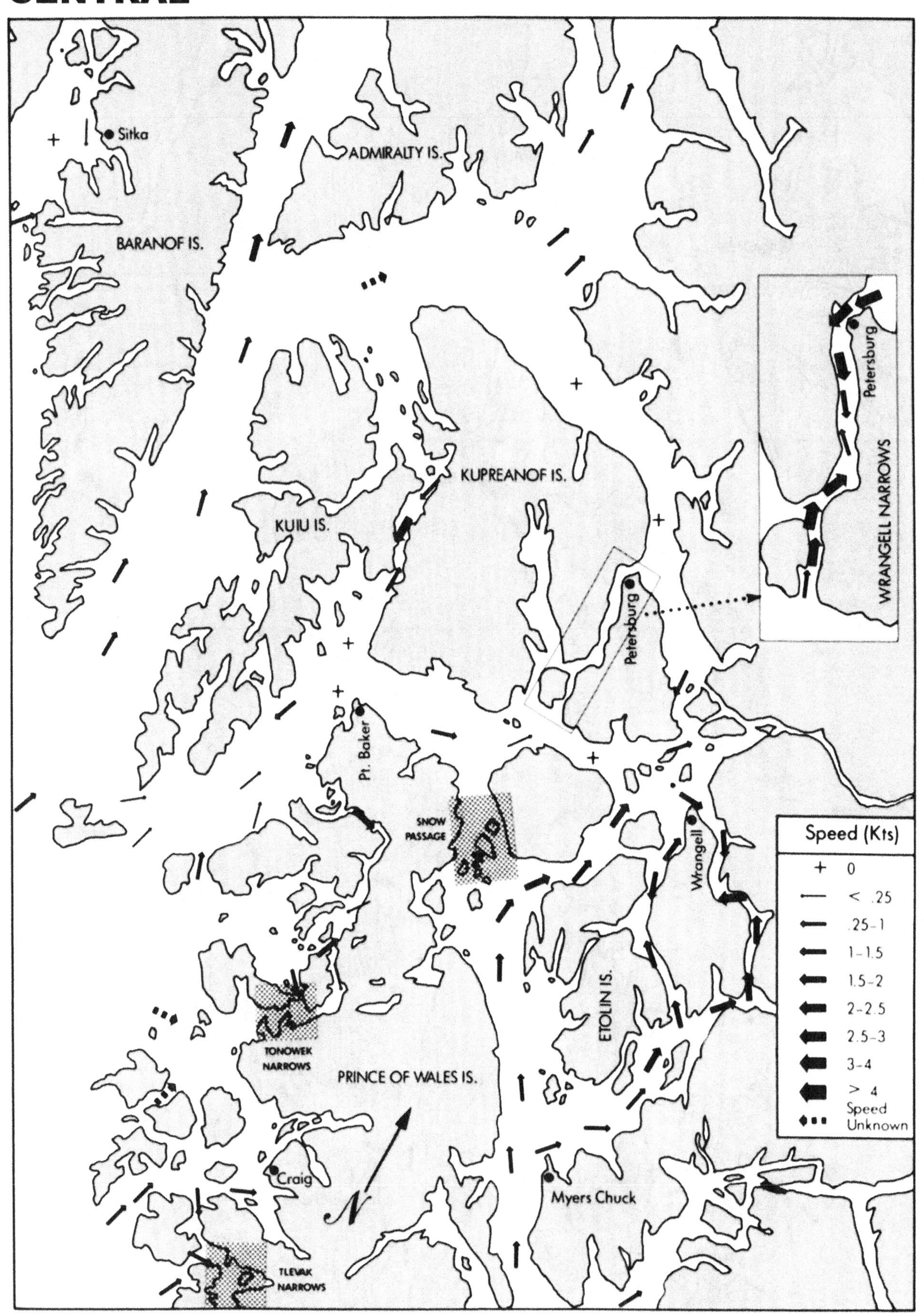

Sitka
ADMIRALTY IS.
BARANOF IS.
KUPREANOF IS.
KUIU IS.
Petersburg
WRANGELL NARROWS
Pt. Baker
SNOW PASSAGE
Wrangell
ETOLIN IS.
TONOWEK NARROWS
PRINCE OF WALES IS.
Craig
Myers Chuck
TLEVAK NARROWS
N
Speed (Kts)
0
< .25
.25–1
1–1.5
1.5–2
2–2.5
2.5–3
3–4
> 4
Speed Unknown

Shaded Areas: See SELECTED CHANNELS

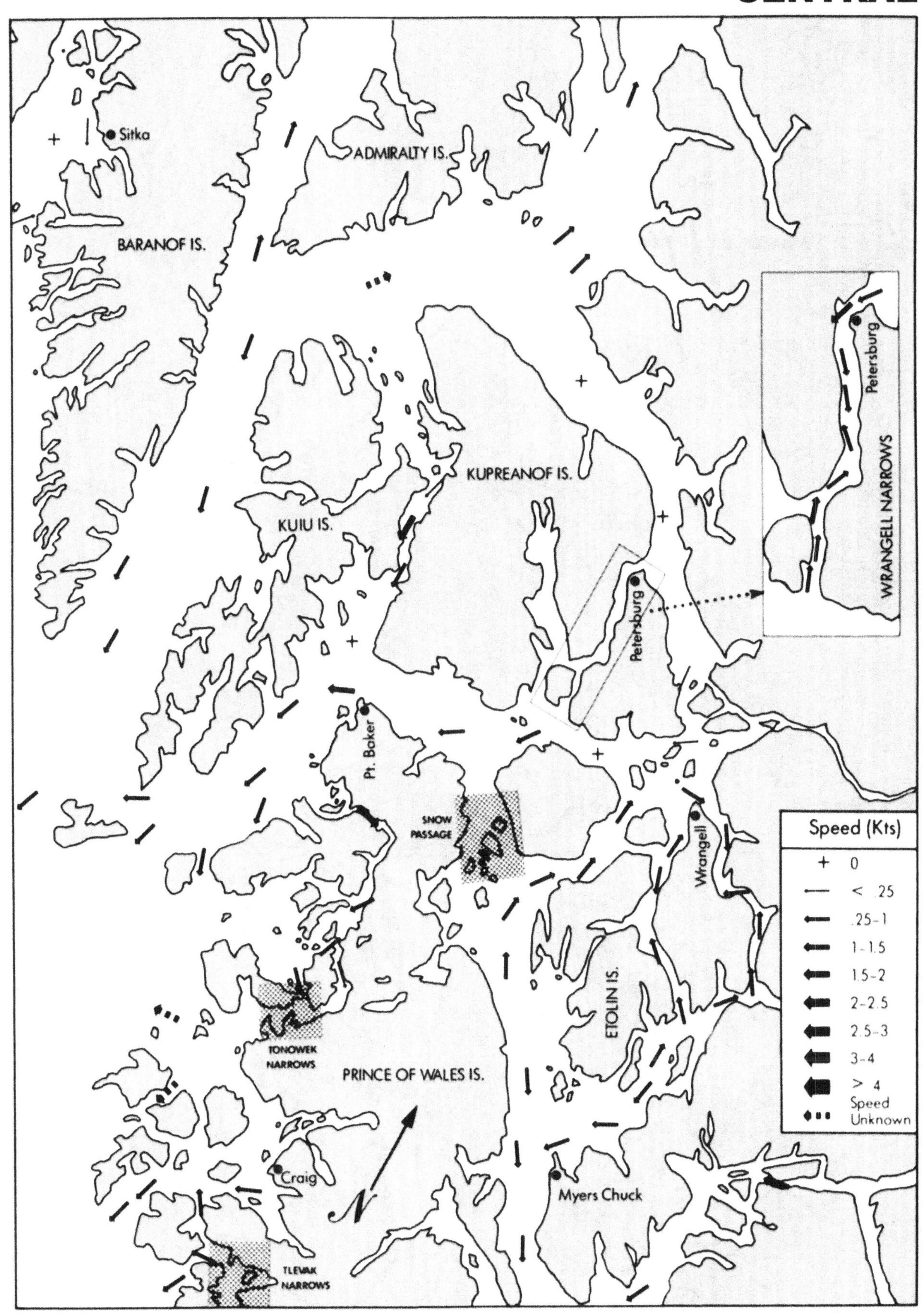
Sitka
ADMIRALTY IS.
BARANOF IS.
KUPREANOF IS.
KUIU IS.
Petersburg
WRANGELL NARROWS
Pt. Baker
SNOW PASSAGE
Wrangell
ETOLIN IS.
TONOWEK NARROWS
PRINCE OF WALES IS.
Craig
Myers Chuck
TLEVAK NARROWS
N
Speed (Kts)
0
< .25
.25-1
1-1.5
1.5-2
2-2.5
2.5-3
3-4
> 4
Speed Unknown

Shaded Areas: See SELECTED CHANNELS

CENTRAL

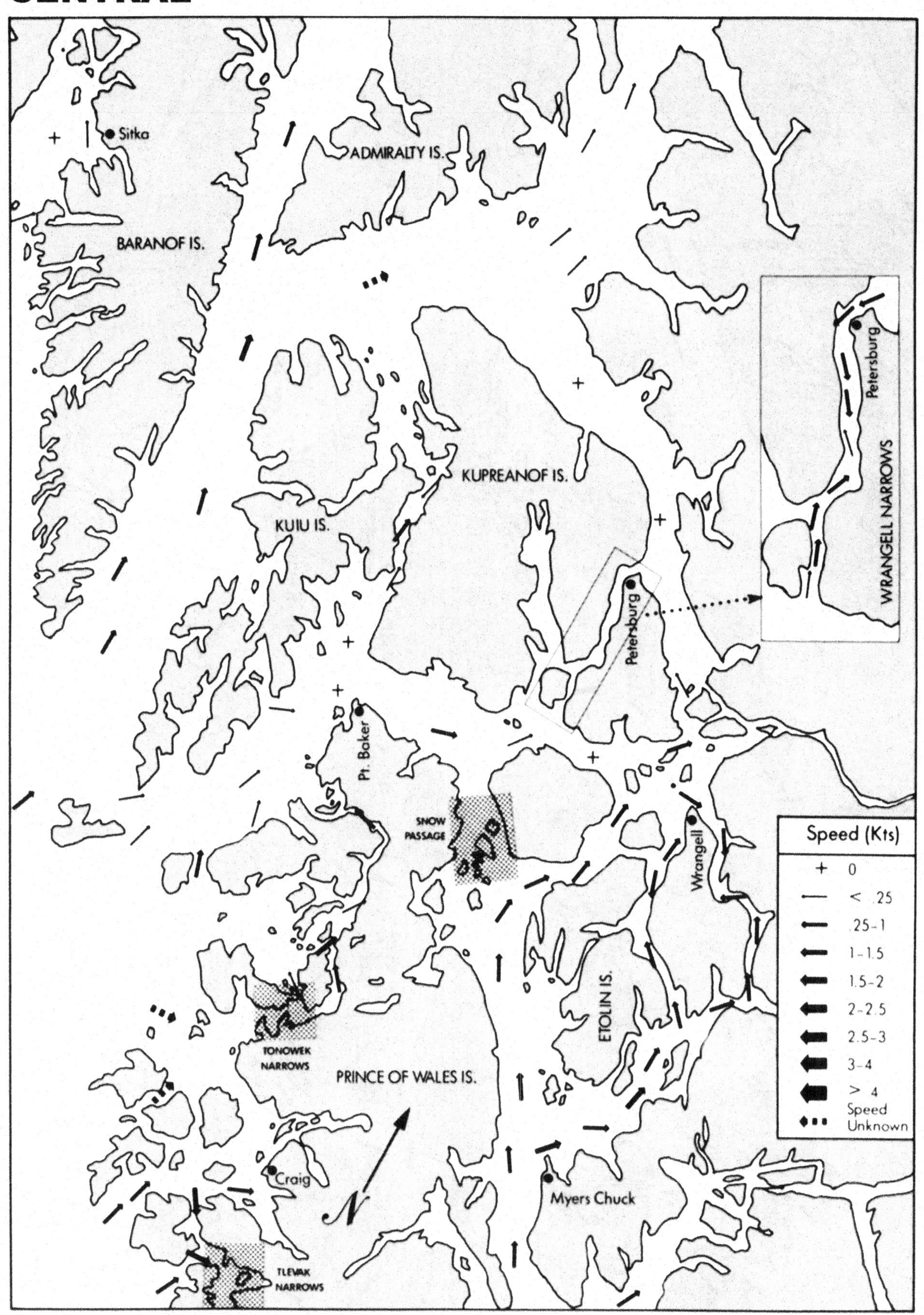

Sitka
ADMIRALTY IS.
BARANOF IS.
KUPREANOF IS.
KUIU IS.
Petersburg
WRANGELL NARROWS
Pt. Baker
SNOW PASSAGE
Wrangell
Speed (Kts)
0
< .25
.25-1
1-1.5
1.5-2
2-2.5
2.5-3
3-4
> 4
Speed Unknown
ETOLIN IS.
TONOWEK NARROWS
PRINCE OF WALES IS.
Craig
N
Myers Chuck
TLEVAK NARROWS

Shaded Areas: See SELECTED CHANNELS

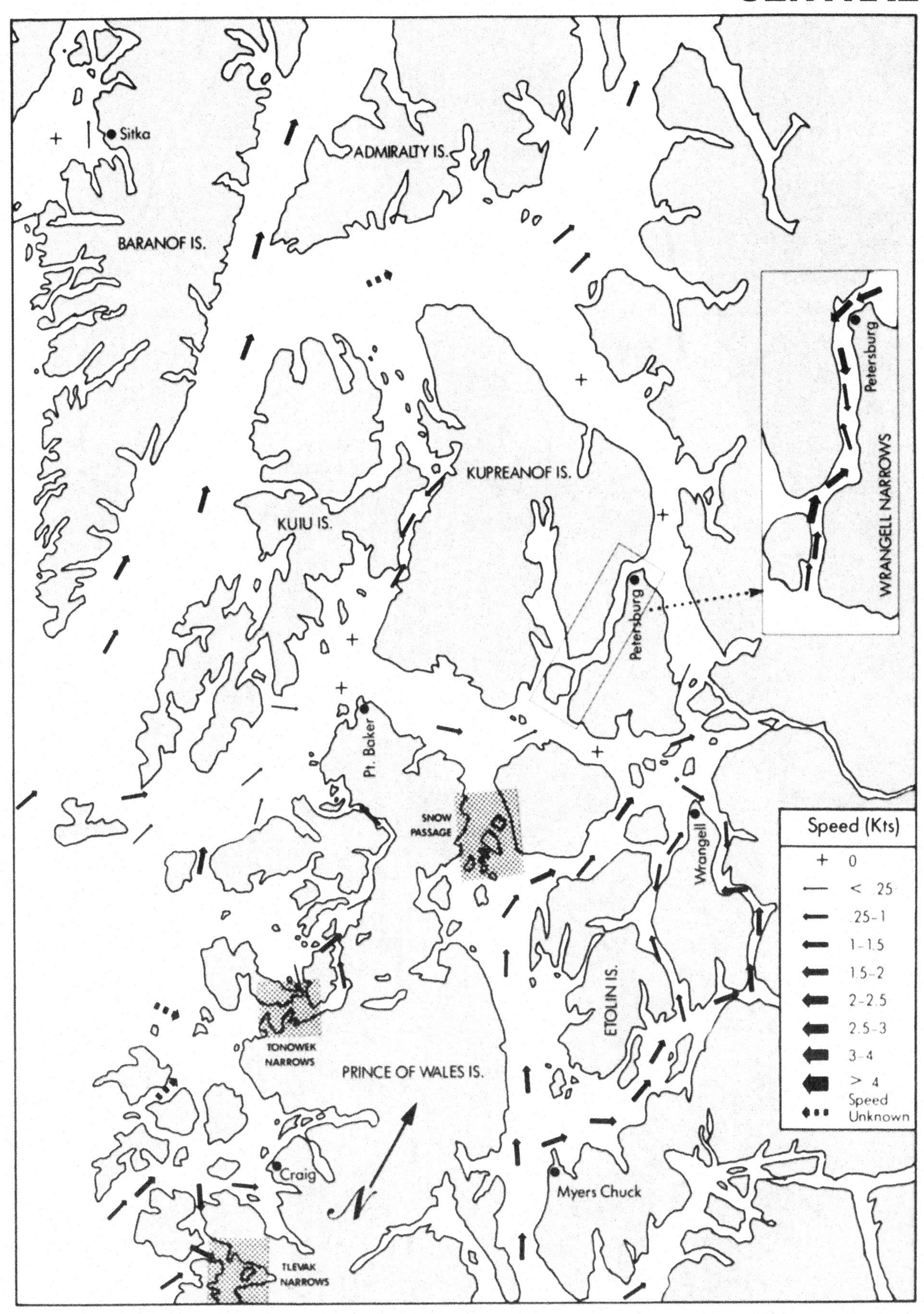
Sitka
ADMIRALTY IS.
BARANOF IS.
KUPREANOF IS.
KUIU IS.
Petersburg
WRANGELL NARROWS
Pt. Baker
SNOW PASSAGE
Wrangell
Speed (Kts)
0
< .25
.25-1
1-1.5
1.5-2
2-2.5
2.5-3
3-4
> 4
Speed Unknown
ETOLIN IS.
TONOWEK NARROWS
PRINCE OF WALES IS.
Craig
Myers Chuck
N
TLEVAK NARROWS

Shaded Areas: See SELECTED CHANNELS

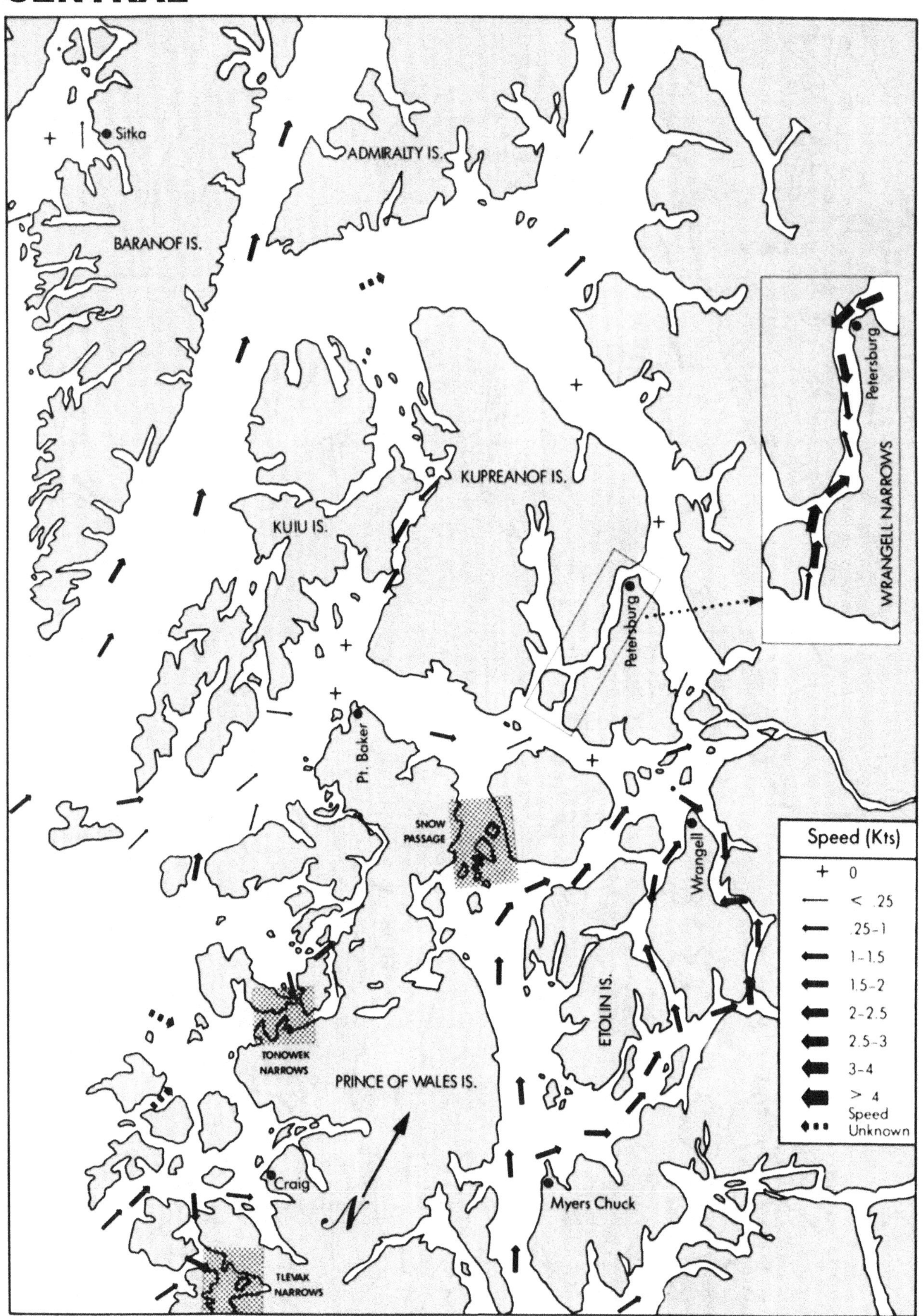
Sitka
ADMIRALTY IS.
BARANOF IS.
KUPREANOF IS.
KUIU IS.
Petersburg
WRANGELL NARROWS
Pt. Baker
SNOW PASSAGE
Wrangell
ETOLIN IS.
TONOWEK NARROWS
PRINCE OF WALES IS.
Craig
Myers Chuck
TLEVAK NARROWS
N
Speed (Kts)
0
< .25
.25-1
1-1.5
1.5-2
2-2.5
2.5-3
3-4
> 4
Speed Unknown

Shaded Areas: See SELECTED CHANNELS

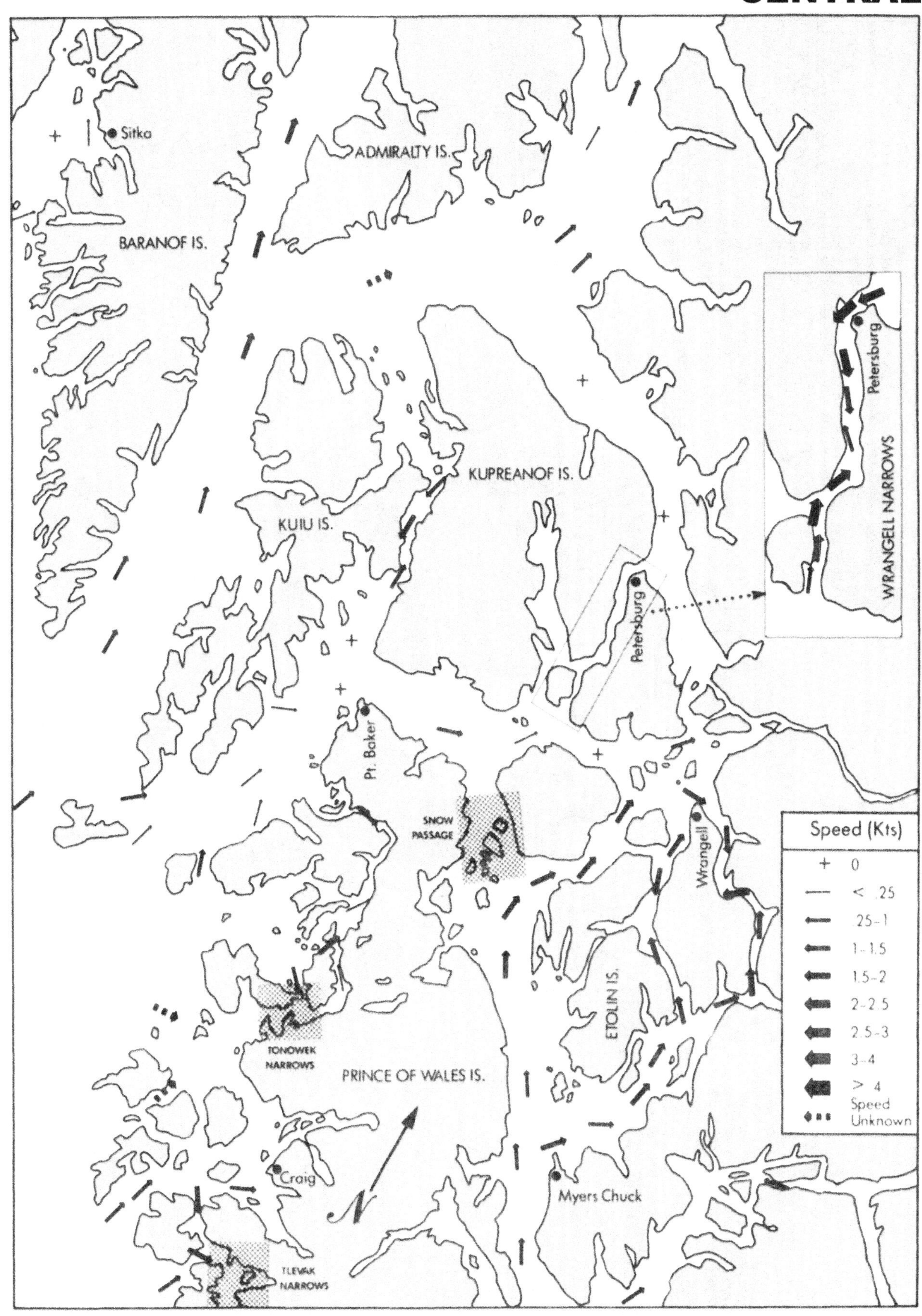
Sitka
ADMIRALTY IS.
BARANOF IS.
KUPREANOF IS.
KUIU IS.
Petersburg
WRANGELL NARROWS
Pt. Baker
SNOW PASSAGE
Wrangell
Speed (Kts)
0
< .25
.25-1
1-1.5
1.5-2
2-2.5
2.5-3
3-4
> 4
Speed Unknown
ETOLIN IS.
TONOWEK NARROWS
PRINCE OF WALES IS.
Craig
Myers Chuck
N
TLEVAK NARROWS

Shaded Areas: See SELECTED CHANNELS

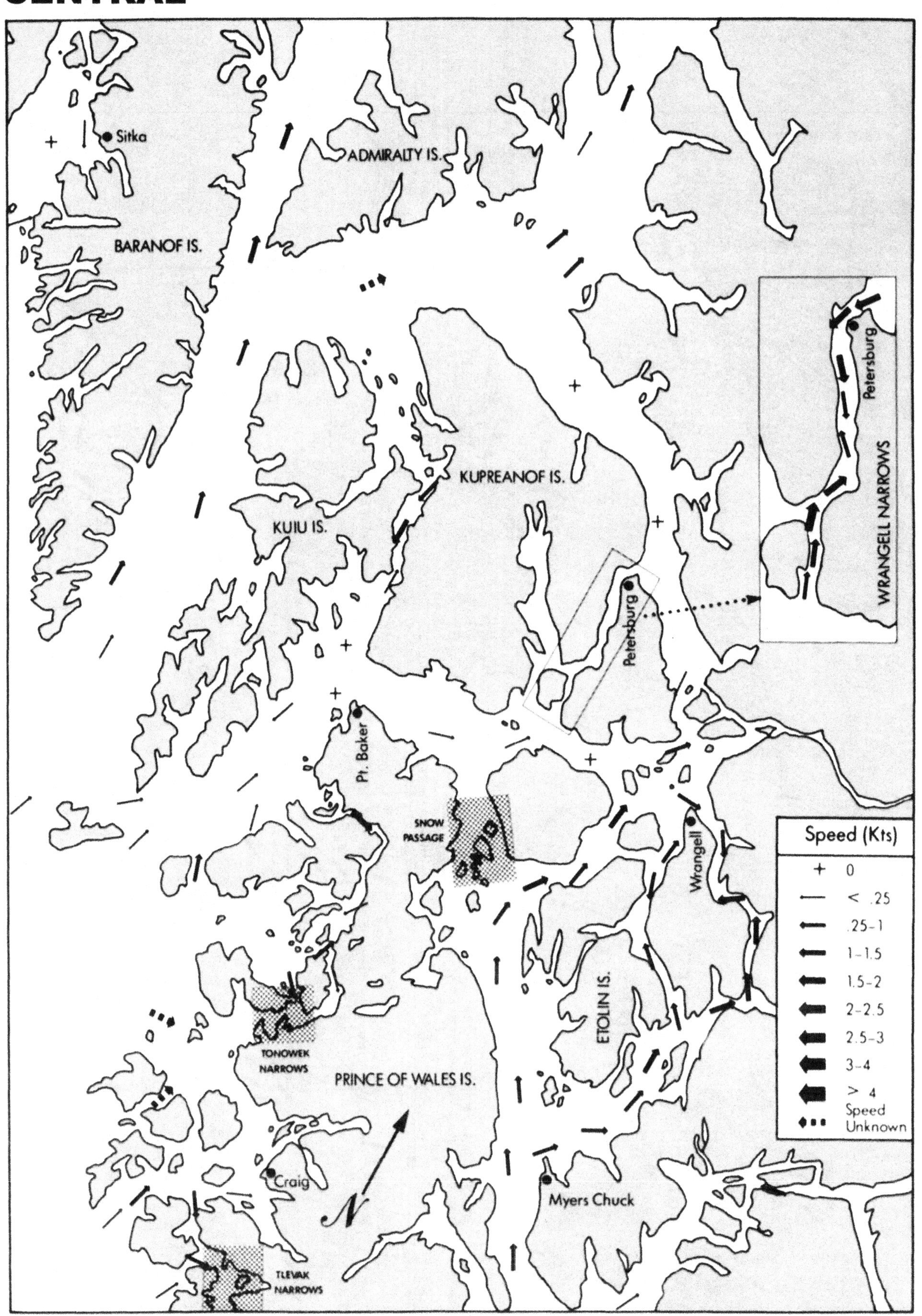
Sitka
ADMIRALTY IS.
BARANOF IS.
KUPREANOF IS.
KUIU IS.
Petersburg
Petersburg
WRANGELL NARROWS
Pt. Baker
SNOW PASSAGE
Wrangell
ETOLIN IS.
TONOWEK NARROWS
PRINCE OF WALES IS.
Craig
N
Myers Chuck
TLEVAK NARROWS
Speed (Kts)
0
< .25
.25-1
1-1.5
1.5-2
2-2.5
2.5-3
3-4
> 4
Speed Unknown

Shaded Areas: See SELECTED CHANNELS

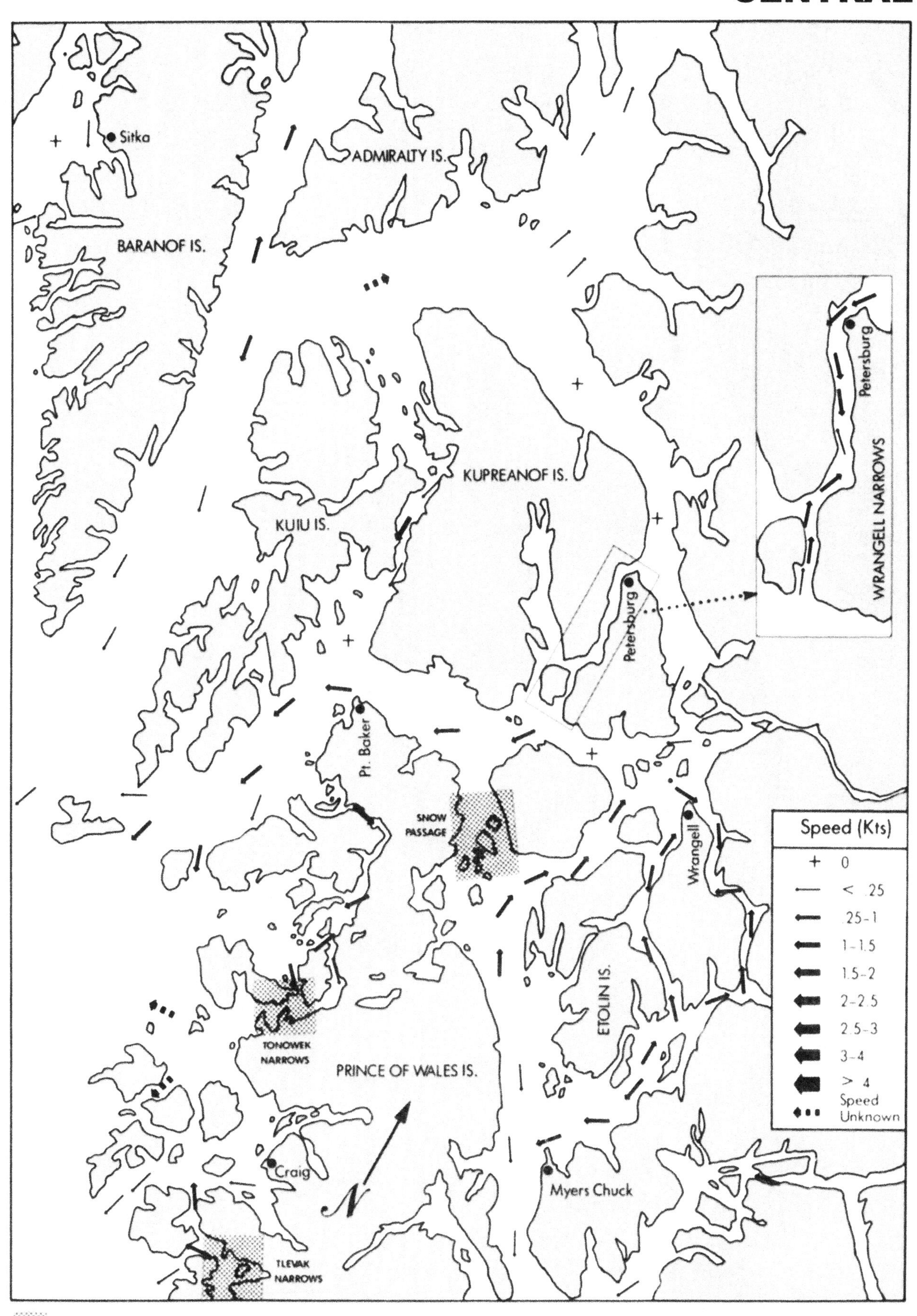
Sitka
ADMIRALTY IS.
BARANOF IS.
KUPREANOF IS.
KUIU IS.
Petersburg
WRANGELL NARROWS
Pt. Baker
SNOW PASSAGE
Wrangell
ETOLIN IS.
TONOWEK NARROWS
PRINCE OF WALES IS.
Craig
Myers Chuck
TLEVAK NARROWS
Speed (Kts)
0
< .25
.25-1
1-1.5
1.5-2
2-2.5
2.5-3
3-4
> 4
Speed Unknown

Shaded Areas: See SELECTED CHANNELS

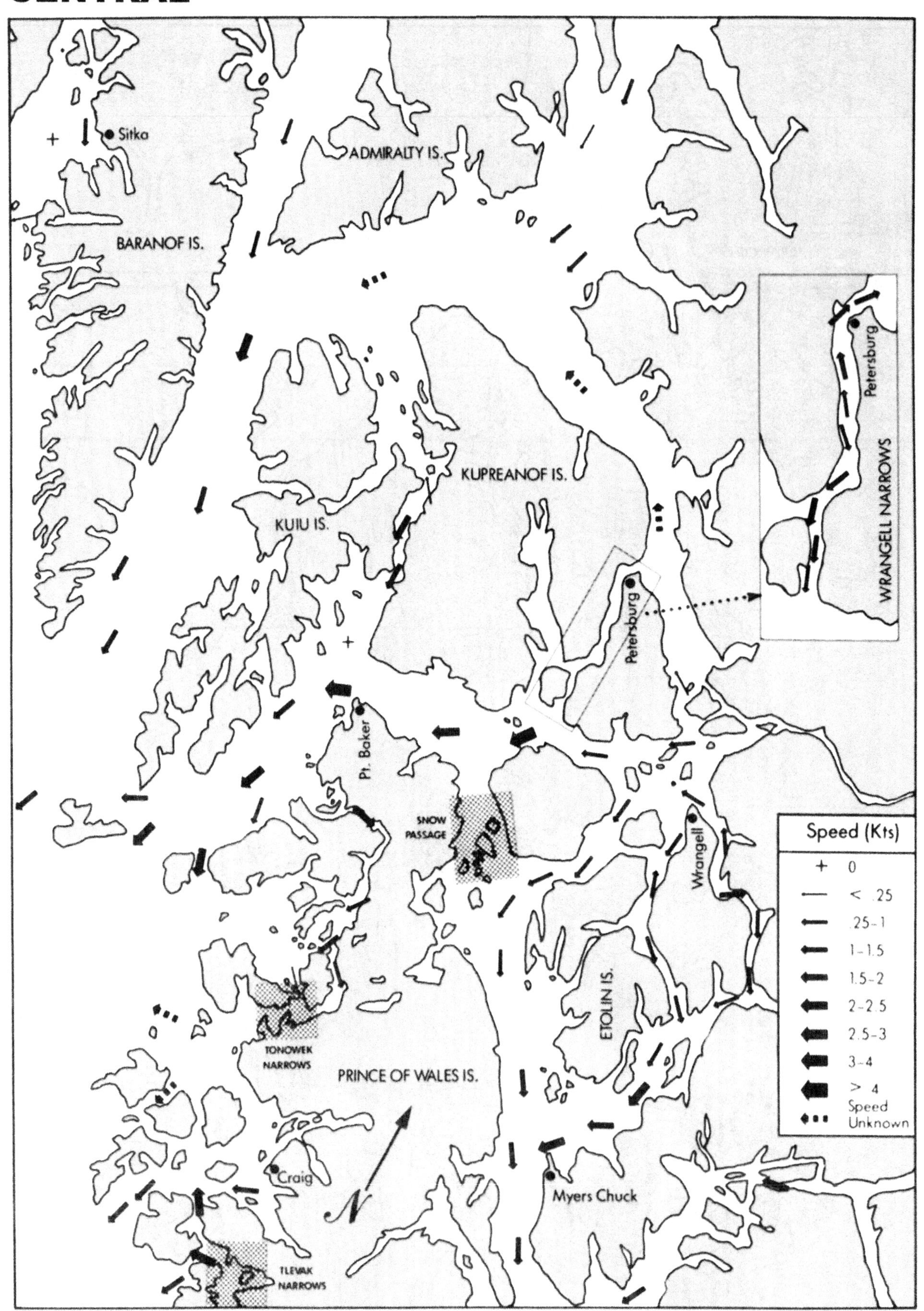
Sitka
ADMIRALTY IS.
BARANOF IS.
KUPREANOF IS.
KUIU IS.
Petersburg
WRANGELL NARROWS
Pt. Baker
SNOW PASSAGE
Wrangell
ETOLIN IS.
TONOWEK NARROWS
PRINCE OF WALES IS.
Craig
Myers Chuck
TLEVAK NARROWS
Speed (Kts)
+ 0
< .25
.25-1
1-1.5
1.5-2
2-2.5
2.5-3
3-4
> 4
Speed Unknown

Shaded Areas: See SELECTED CHANNELS

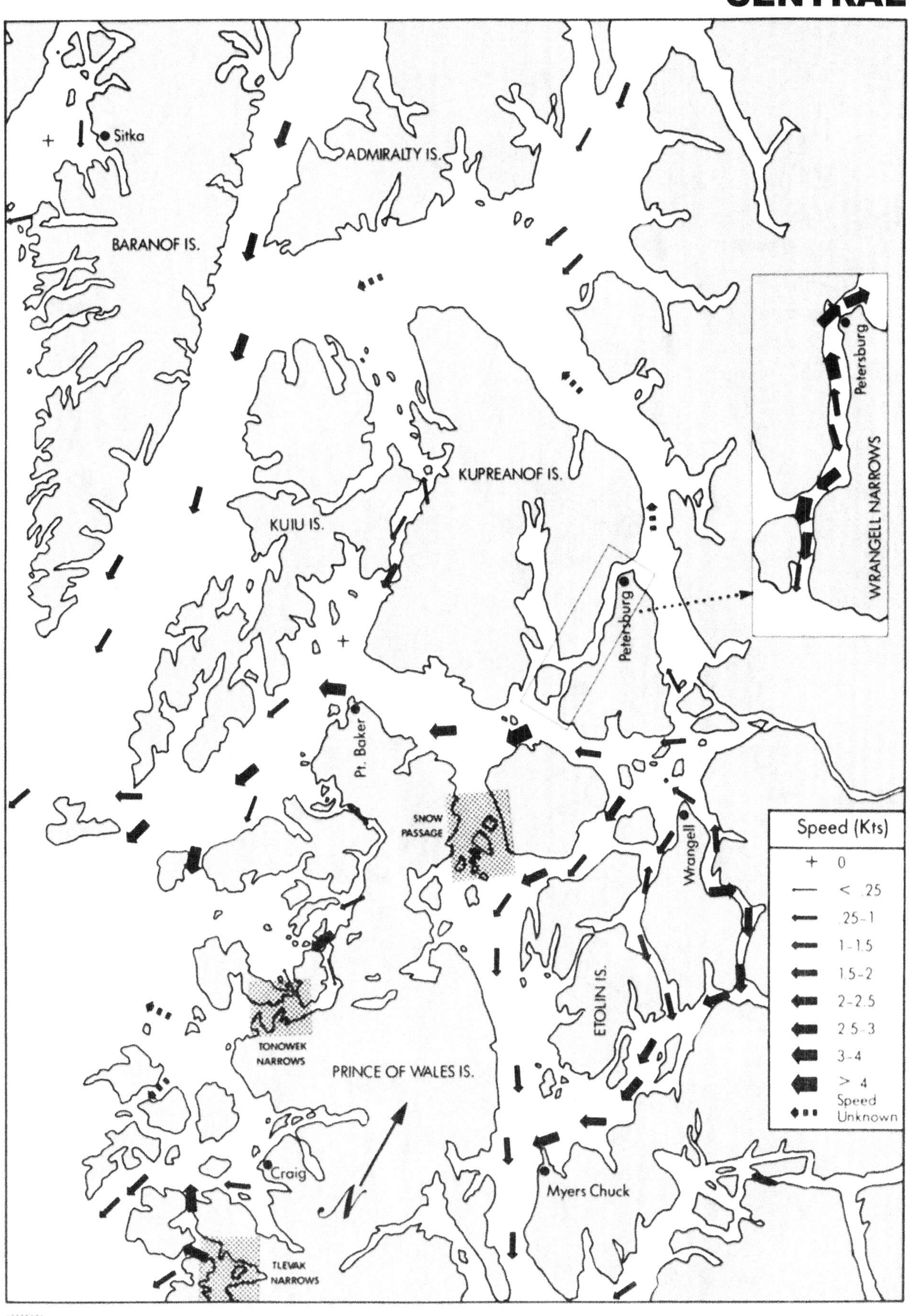
Sitka
ADMIRALTY IS.
BARANOF IS.
KUPREANOF IS.
KUIU IS.
Petersburg
WRANGELL NARROWS
Pt. Baker
SNOW PASSAGE
Wrangell
ETOLIN IS.
TONOWEK NARROWS
PRINCE OF WALES IS.
Craig
Myers Chuck
TLEVAK NARROWS
Speed (Kts)
0
< .25
.25-1
1-1.5
1.5-2
2-2.5
2.5-3
3-4
> 4
Speed Unknown

Shaded Areas: See SELECTED CHANNELS

CENTRAL

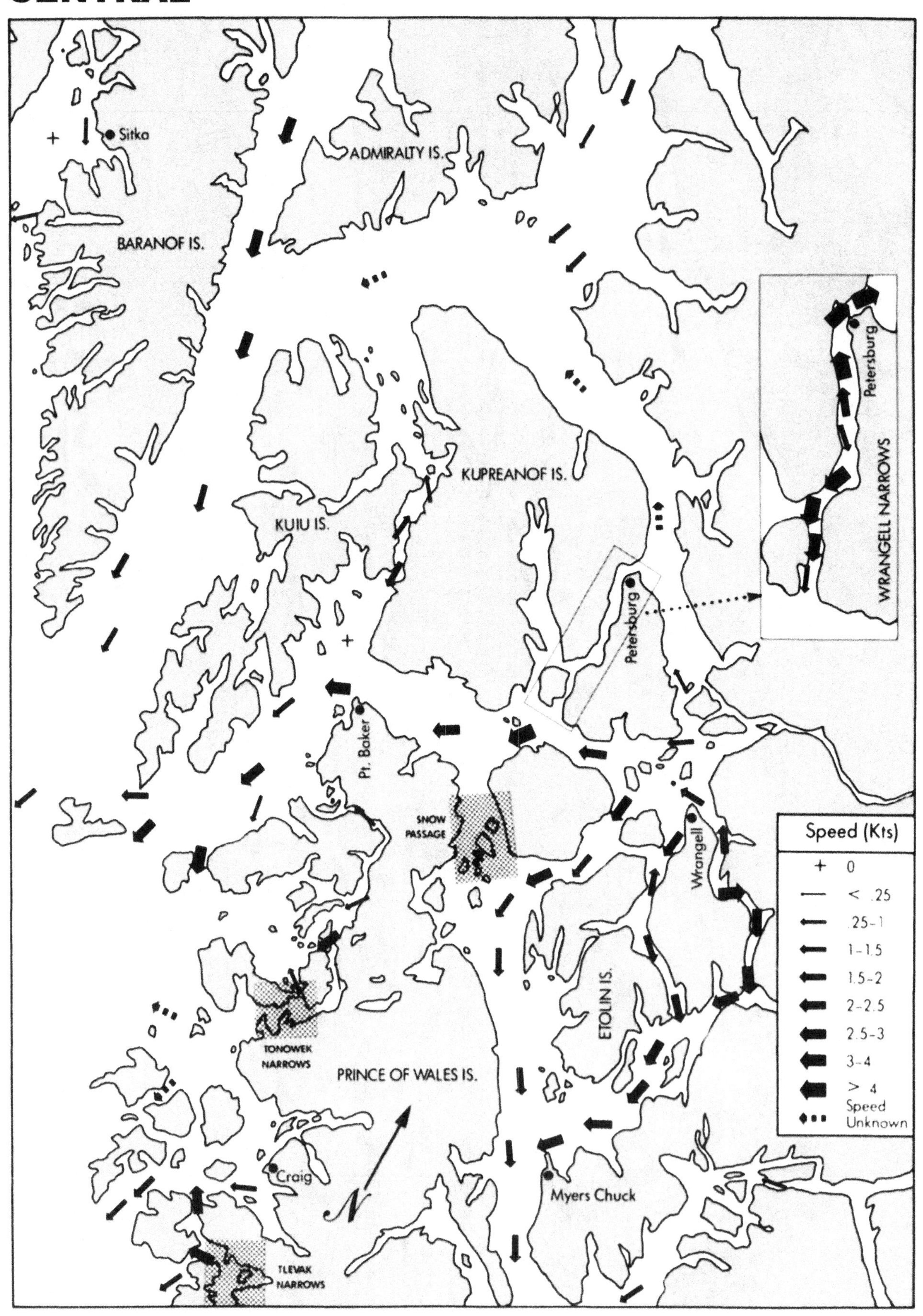

Sitka
ADMIRALTY IS.
BARANOF IS.
KUPREANOF IS.
KUIU IS.
Petersburg
Petersburg
WRANGELL NARROWS
Pt. Baker
SNOW PASSAGE
Wrangell
Speed (Kts)
0
< .25
.25-1
1-1.5
1.5-2
2-2.5
2.5-3
3-4
> 4
Speed Unknown
ETOLIN IS.
TONOWEK NARROWS
PRINCE OF WALES IS.
Craig
N
Myers Chuck
TLEVAK NARROWS

Shaded Areas: See SELECTED CHANNELS

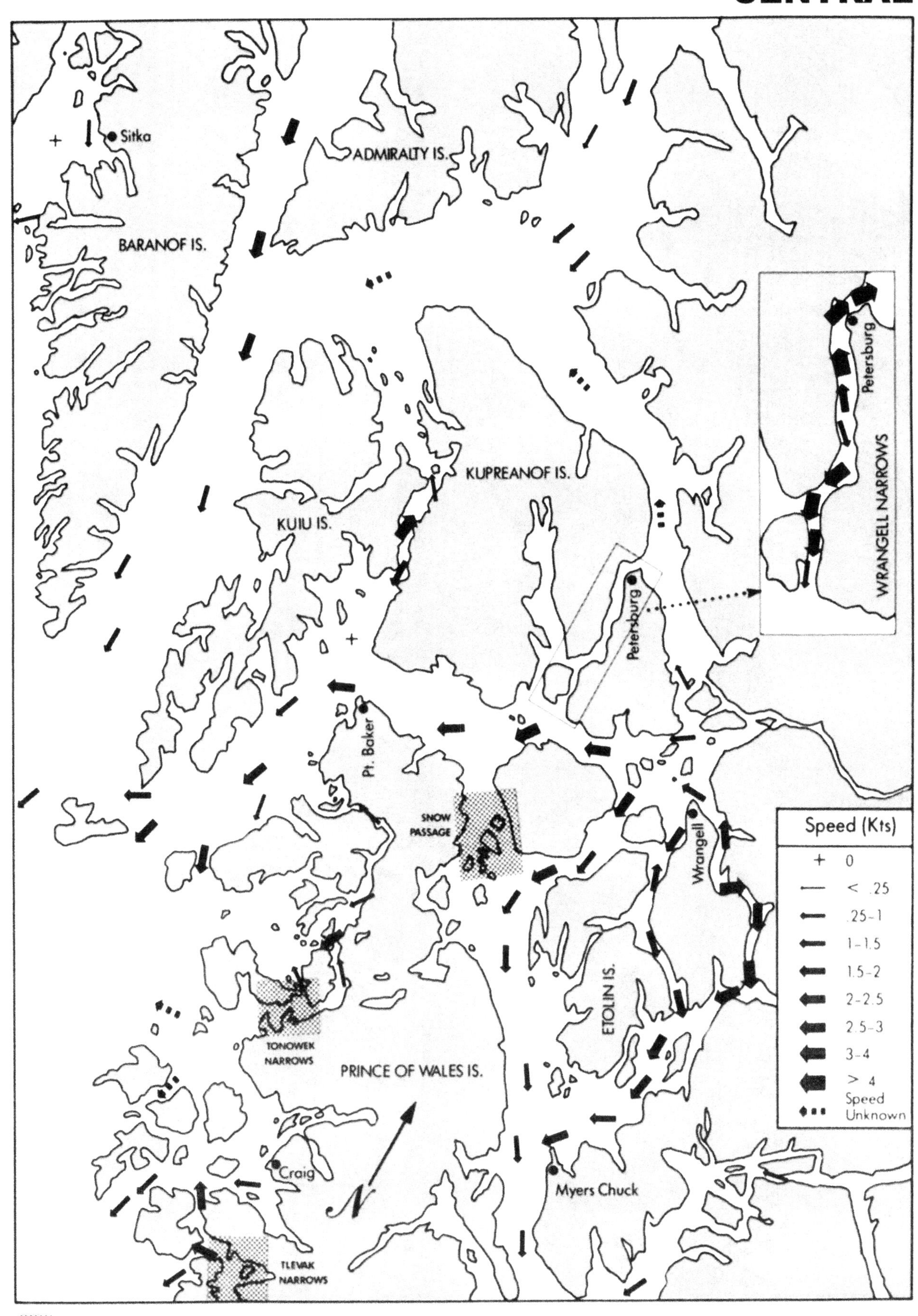
Sitka
ADMIRALTY IS.
BARANOF IS.
KUPREANOF IS.
KUIU IS.
Petersburg
Petersburg
WRANGELL NARROWS
Pt. Baker
SNOW PASSAGE
Wrangell
ETOLIN IS.
TONOWEK NARROWS
PRINCE OF WALES IS.
Craig
Myers Chuck
TLEVAK NARROWS
Speed (Kts)
+ 0
< .25
.25-1
1-1.5
1.5-2
2-2.5
2.5-3
3-4
> 4
Speed Unknown

Shaded Areas: See SELECTED CHANNELS

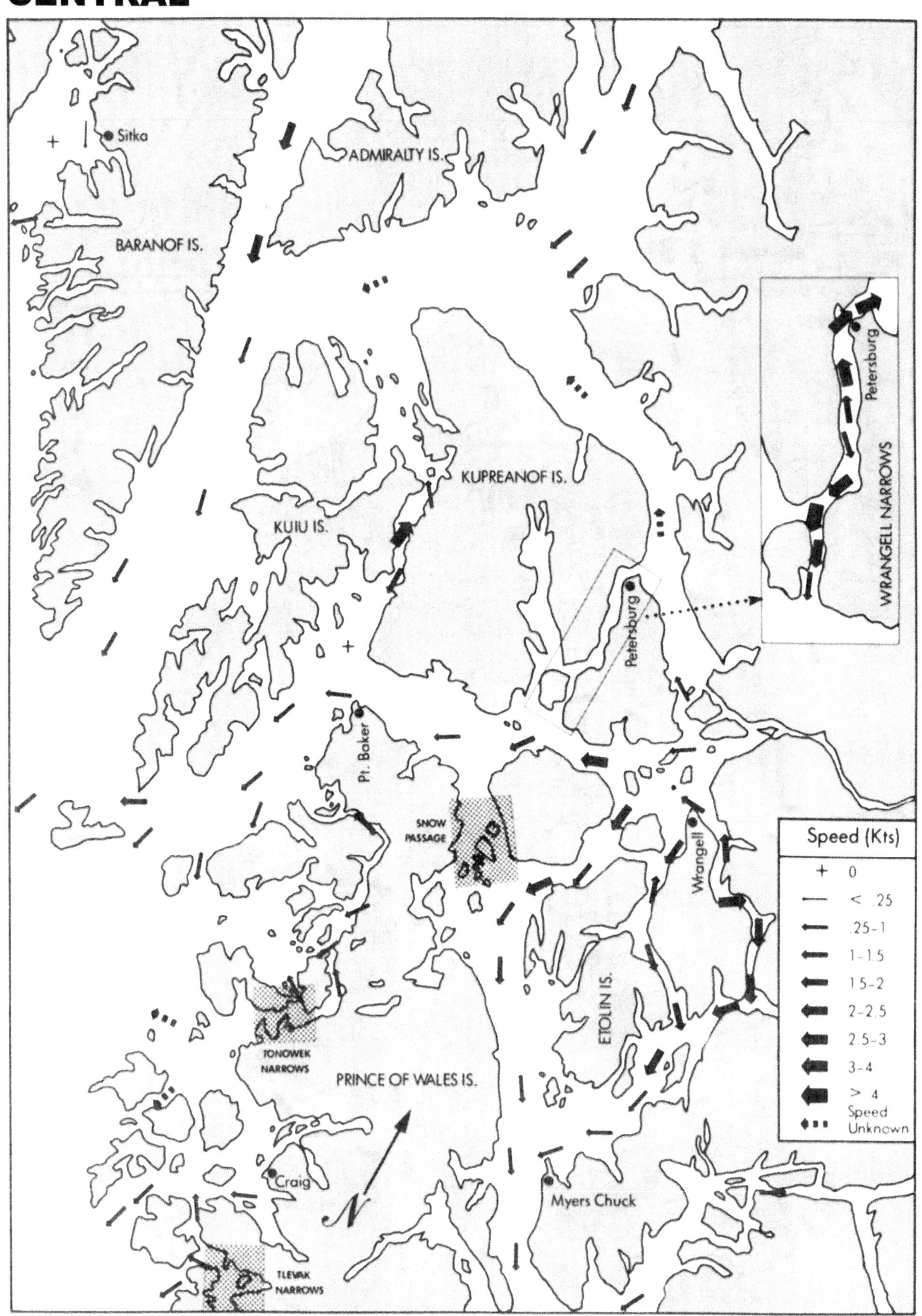
Sitka
ADMIRALTY IS.
BARANOF IS.
KUPREANOF IS.
KUIU IS.
Petersburg
WRANGELL NARROWS
Pt. Baker
SNOW PASSAGE
Wrangell
ETOLIN IS.
TONOWEK NARROWS
PRINCE OF WALES IS.
Craig
Myers Chuck
TLEVAK NARROWS
Speed (Kts)
0
< .25
.25-1
1-1.5
1.5-2
2-2.5
2.5-3
3-4
> 4
Speed Unknown

Shaded Areas: See SELECTED CHANNELS

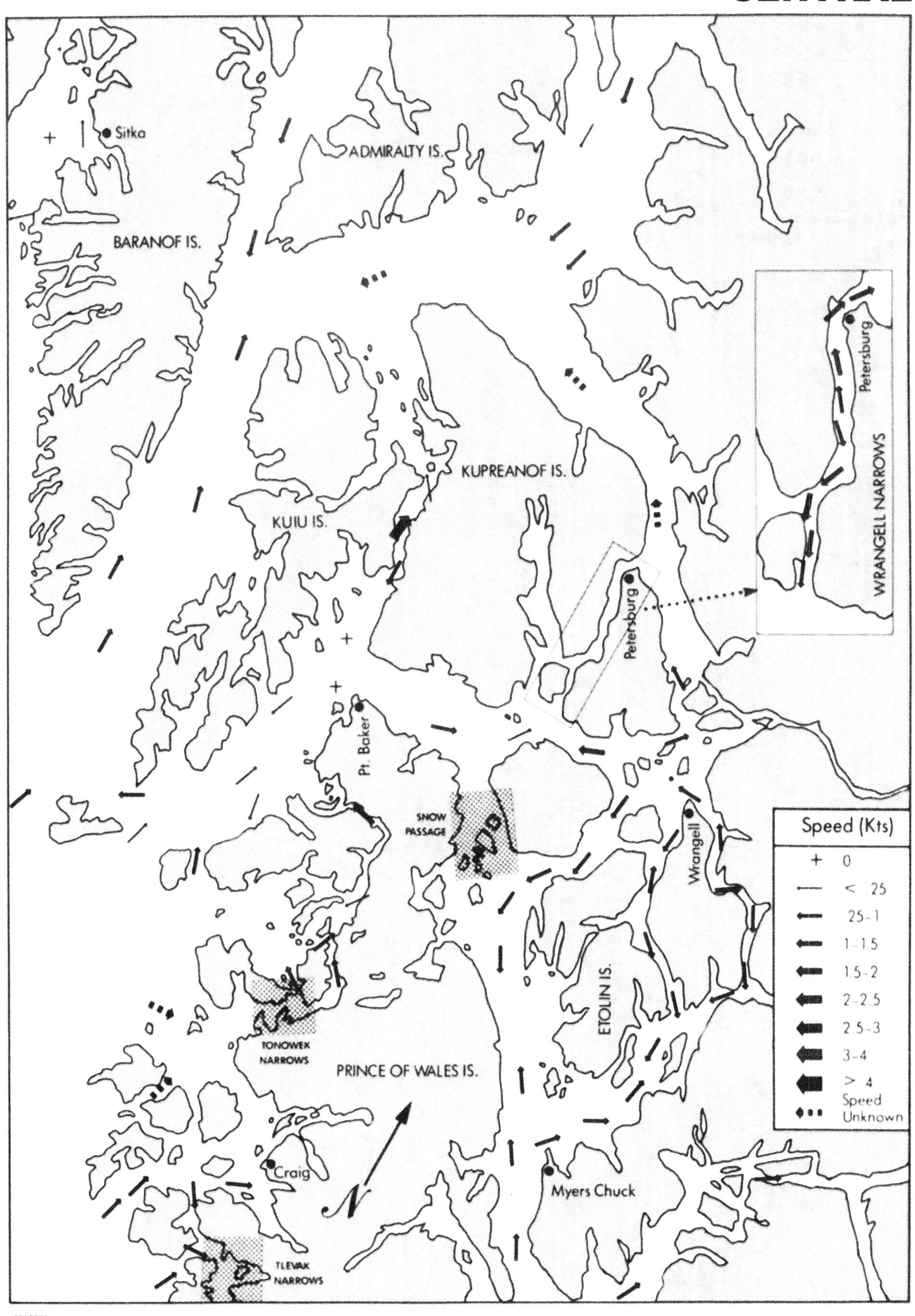

Sitka
ADMIRALTY IS.
BARANOF IS.
KUPREANOF IS.
KUIU IS.
Petersburg
WRANGELL NARROWS
Pt. Baker
SNOW PASSAGE
Wrangell
ETOLIN IS.
TONOWEK NARROWS
PRINCE OF WALES IS.
Craig
Myers Chuck
TLEVAK NARROWS
Speed (Kts)
0
< .25
.25-1
1-1.5
1.5-2
2-2.5
2.5-3
3-4
> 4
Speed Unknown

Shaded Areas: See SELECTED CHANNELS

CENTRAL

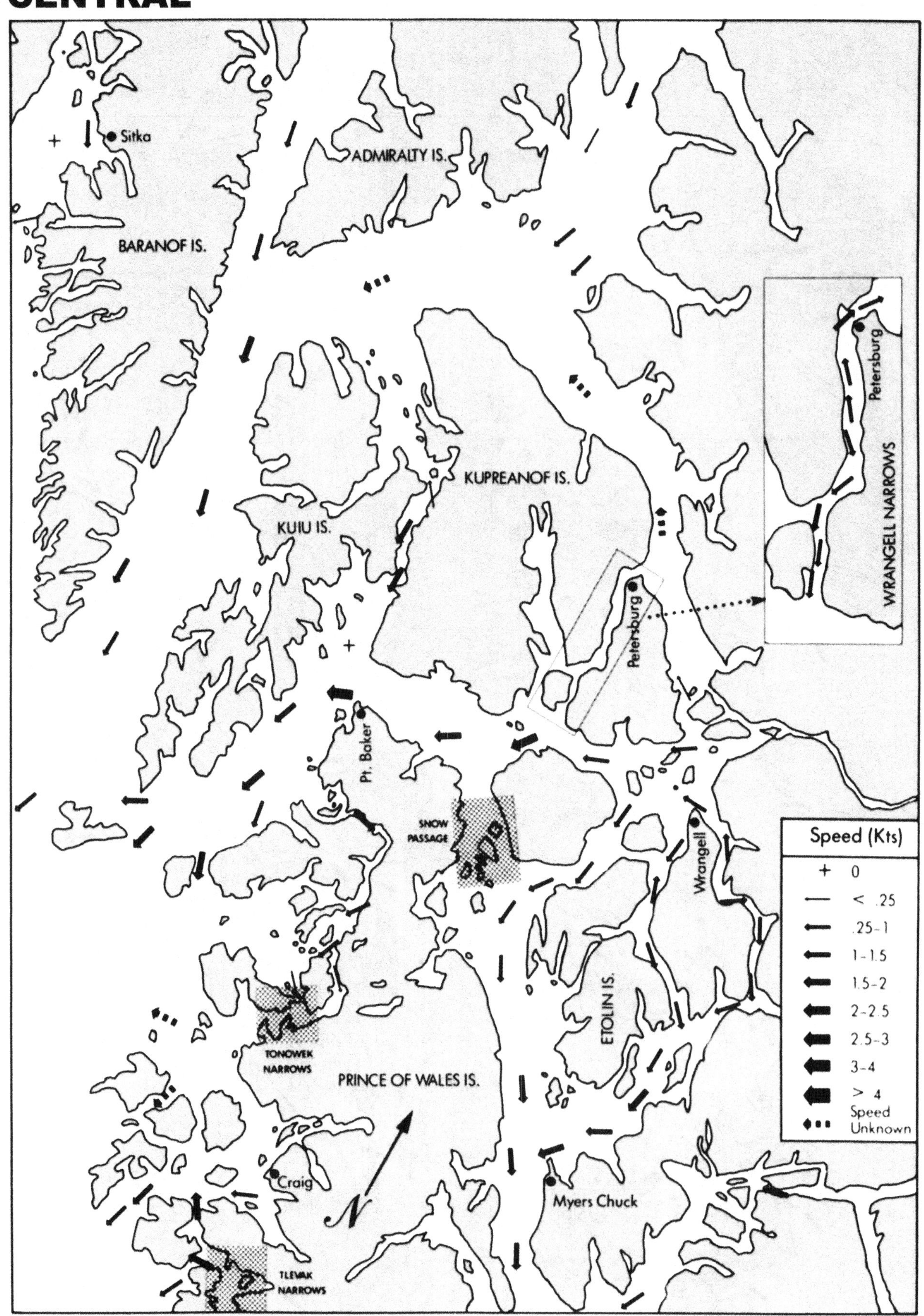

Sitka
ADMIRALTY IS.
BARANOF IS.
KUPREANOF IS.
KUIU IS.
Petersburg
WRANGELL NARROWS
Petersburg
Pt. Baker
SNOW PASSAGE
Wrangell
Speed (Kts)
0
< .25
.25-1
1-1.5
1.5-2
2-2.5
2.5-3
3-4
> 4
Speed Unknown
ETOLIN IS.
TONOWEK NARROWS
PRINCE OF WALES IS.
Craig
N
Myers Chuck
TLEVAK NARROWS

Shaded Areas: See SELECTED CHANNELS

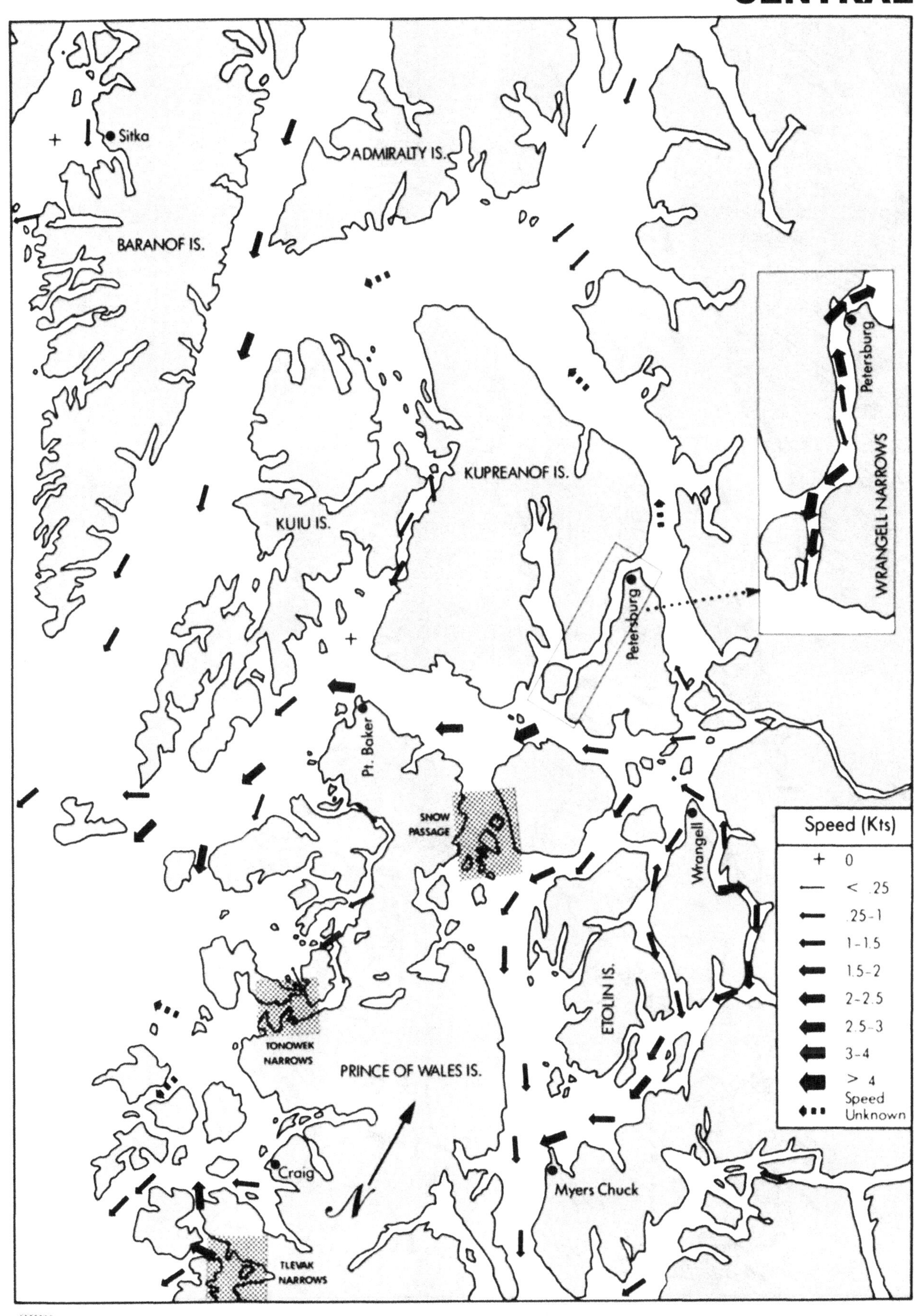
Sitka
ADMIRALTY IS.
BARANOF IS.
KUPREANOF IS.
KUIU IS.
Petersburg
WRANGELL NARROWS
Pt. Baker
SNOW PASSAGE
Wrangell
Speed (Kts)
+ 0
< .25
.25-1
1-1.5
1.5-2
2-2.5
2.5-3
3-4
> 4
Speed Unknown
ETOLIN IS.
TONOWEK NARROWS
PRINCE OF WALES IS.
Craig
Myers Chuck
TLEVAK NARROWS

Shaded Areas: See SELECTED CHANNELS

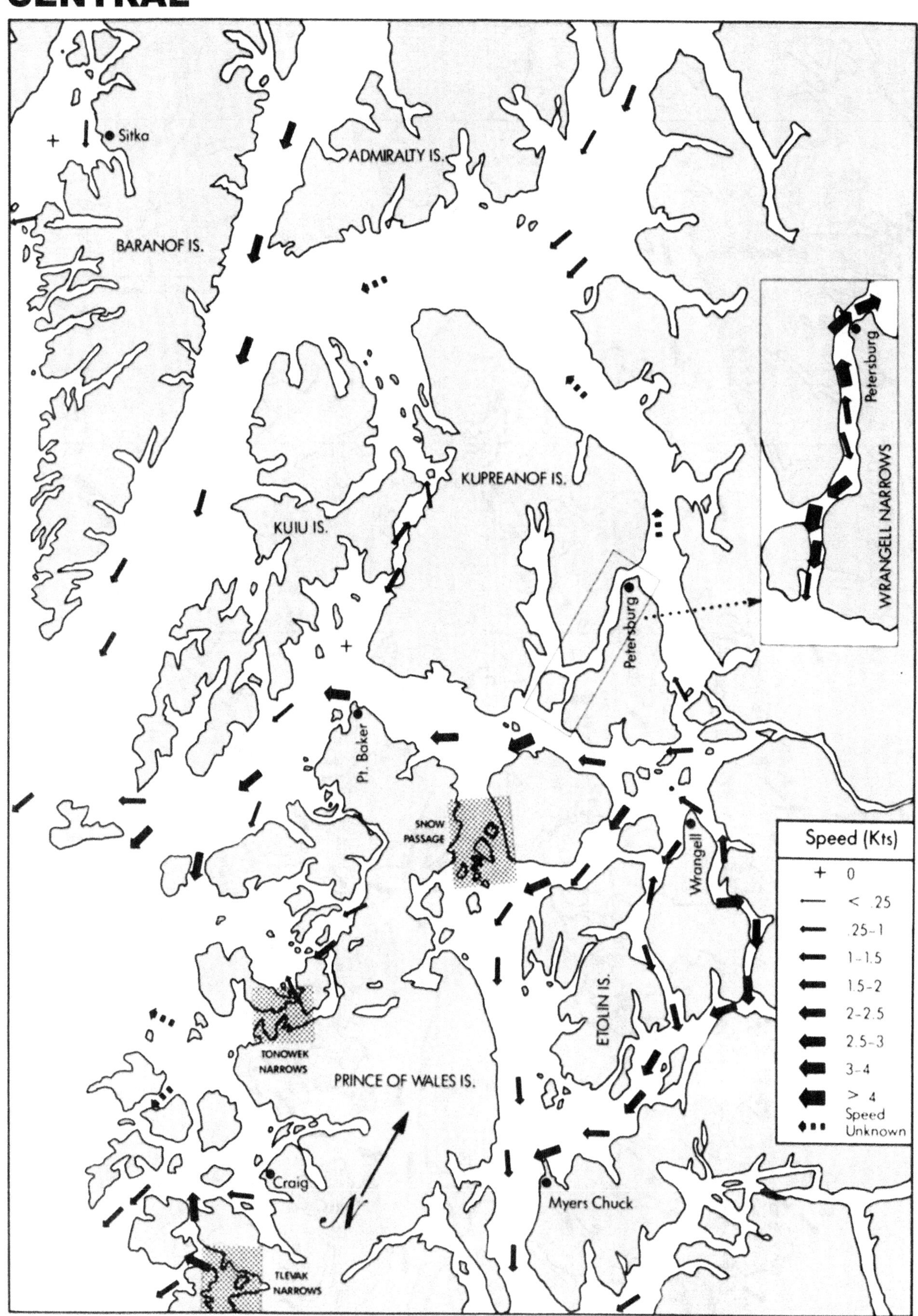
Sitka
ADMIRALTY IS.
BARANOF IS.
KUPREANOF IS.
KUIU IS.
Petersburg
WRANGELL NARROWS
Pt. Baker
SNOW PASSAGE
Wrangell
ETOLIN IS.
TONOWEK NARROWS
PRINCE OF WALES IS.
Craig
Myers Chuck
TLEVAK NARROWS
N
Speed (Kts)
0
< .25
.25–1
1–1.5
1.5–2
2–2.5
2.5–3
3–4
> 4
Speed Unknown

Shaded Areas: See SELECTED CHANNELS

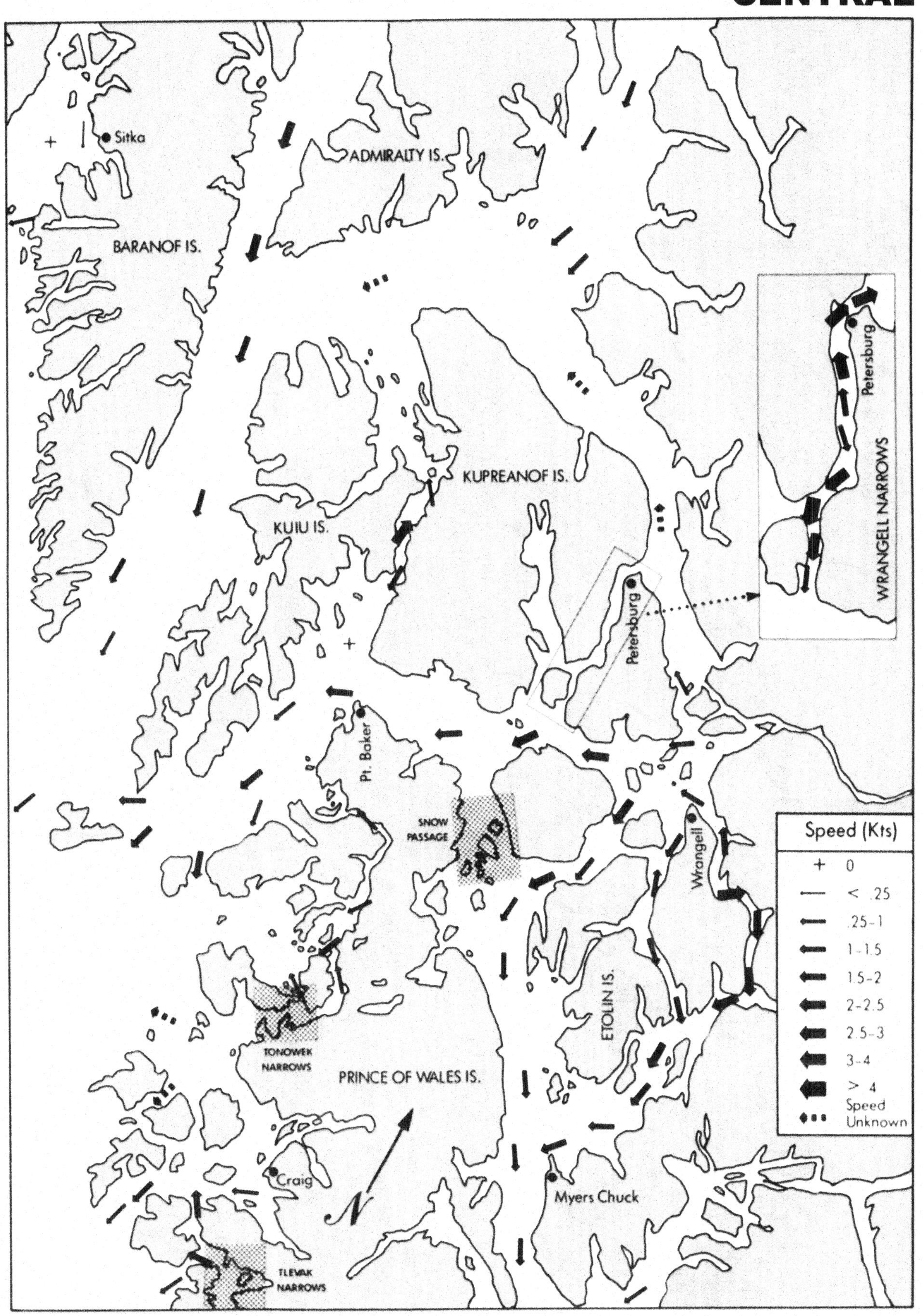
Sitka
ADMIRALTY IS.
BARANOF IS.
KUPREANOF IS.
KUIU IS.
Petersburg
WRANGELL NARROWS
Pt. Baker
SNOW PASSAGE
Wrangell
ETOLIN IS.
TONOWEK NARROWS
PRINCE OF WALES IS.
Craig
Myers Chuck
TLEVAK NARROWS
Speed (Kts)
+ 0
< .25
.25-1
1-1.5
1.5-2
2-2.5
2.5-3
3-4
> 4
Speed Unknown

Shaded Areas: See SELECTED CHANNELS

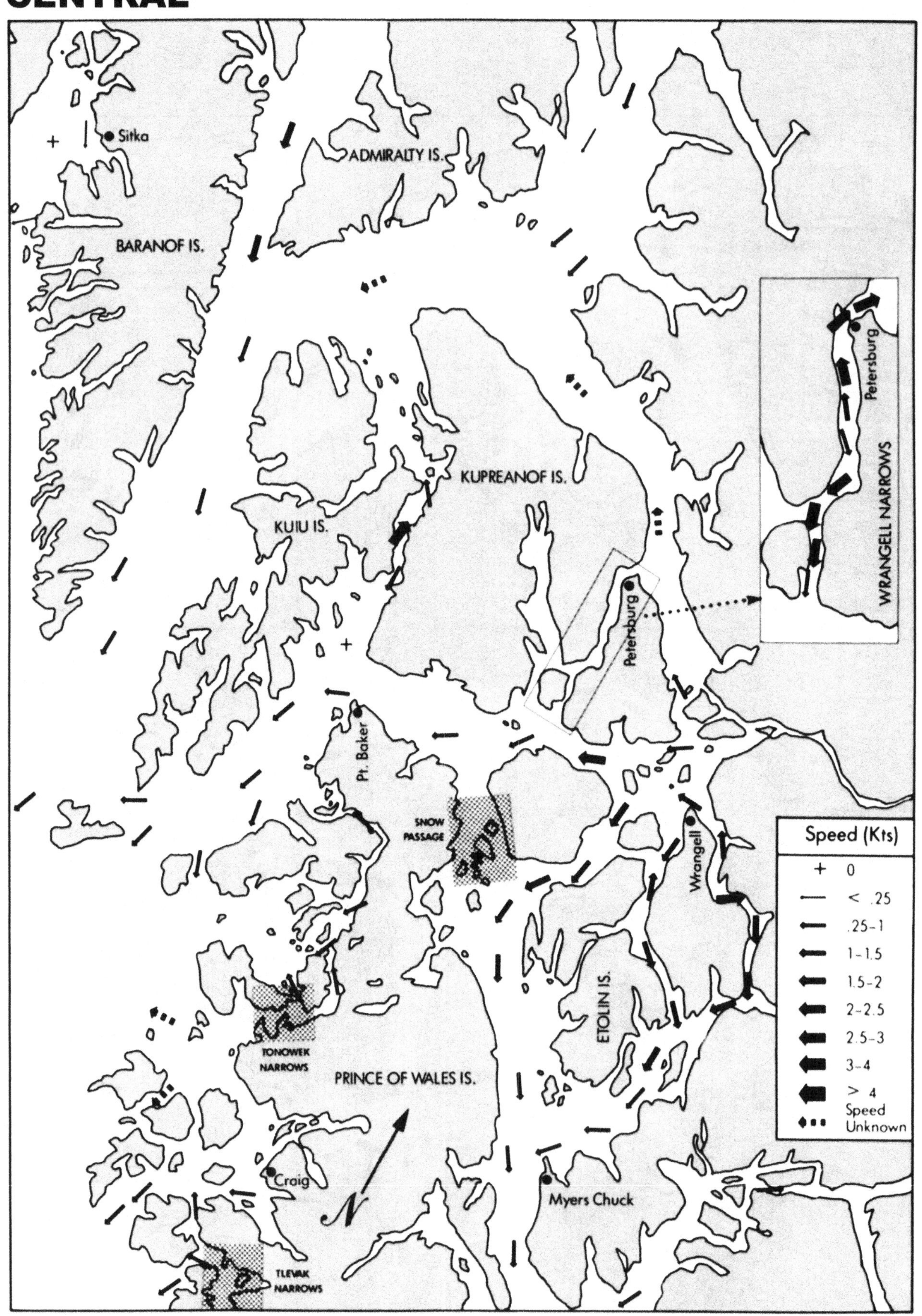
Sitka
ADMIRALTY IS.
BARANOF IS.
KUPREANOF IS.
KUIU IS.
Petersburg
WRANGELL NARROWS
Pt. Baker
SNOW PASSAGE
Wrangell
ETOLIN IS.
TONOWEK NARROWS
PRINCE OF WALES IS.
Craig
Myers Chuck
TLEVAK NARROWS
N
Speed (Kts)
0
< .25
.25-1
1-1.5
1.5-2
2-2.5
2.5-3
3-4
> 4
Speed Unknown

Shaded Areas: See SELECTED CHANNELS

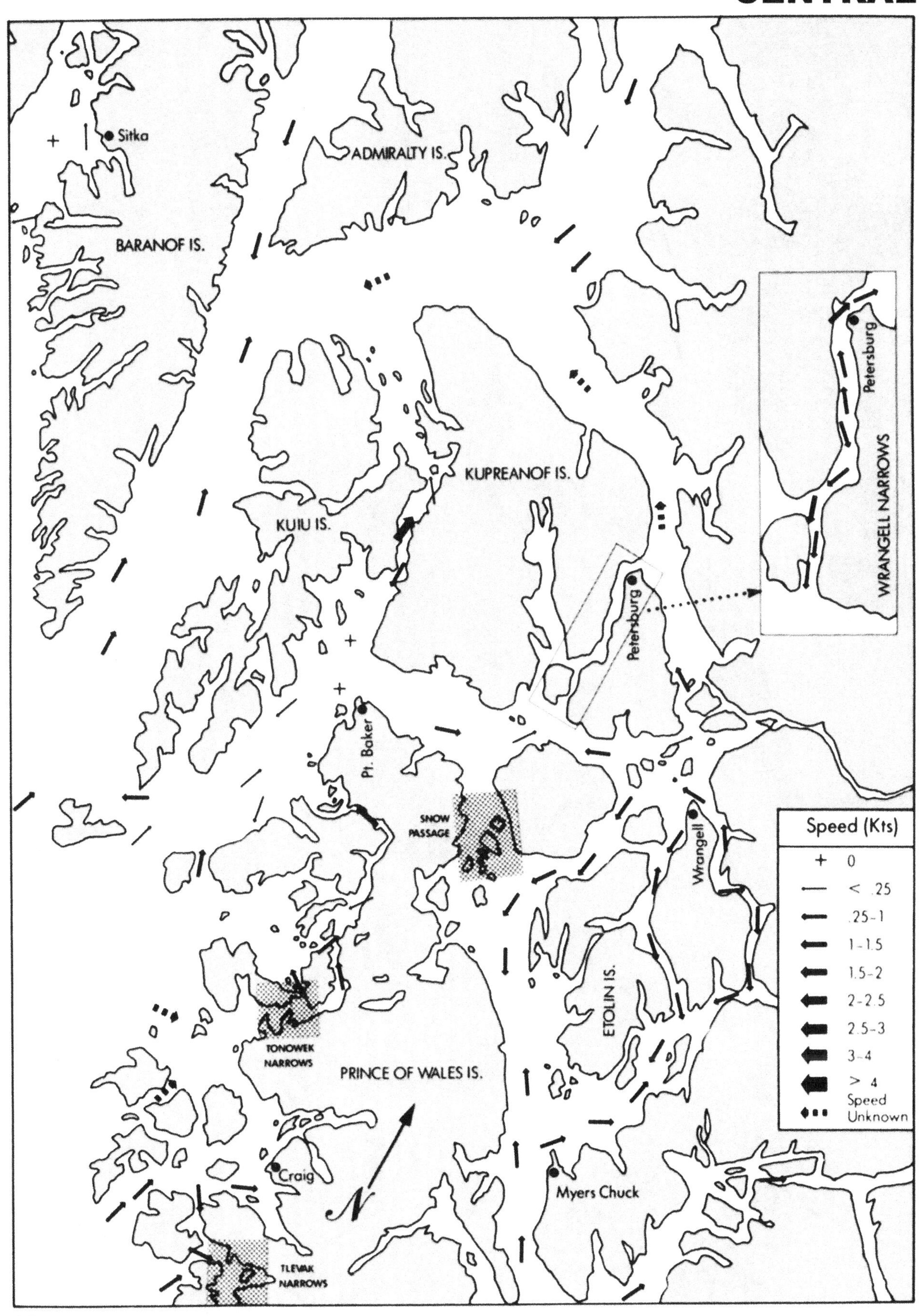
Sitka
ADMIRALTY IS.
BARANOF IS.
KUPREANOF IS.
KUIU IS.
Petersburg
WRANGELL NARROWS
Pt. Baker
SNOW PASSAGE
Wrangell
Speed (Kts)
0
< .25
.25-1
1-1.5
1.5-2
2-2.5
2.5-3
3-4
> 4
Speed Unknown
ETOLIN IS.
TONOWEK NARROWS
PRINCE OF WALES IS.
Craig
Myers Chuck
TLEVAK NARROWS
N

Shaded Areas: See SELECTED CHANNELS

CENTRAL

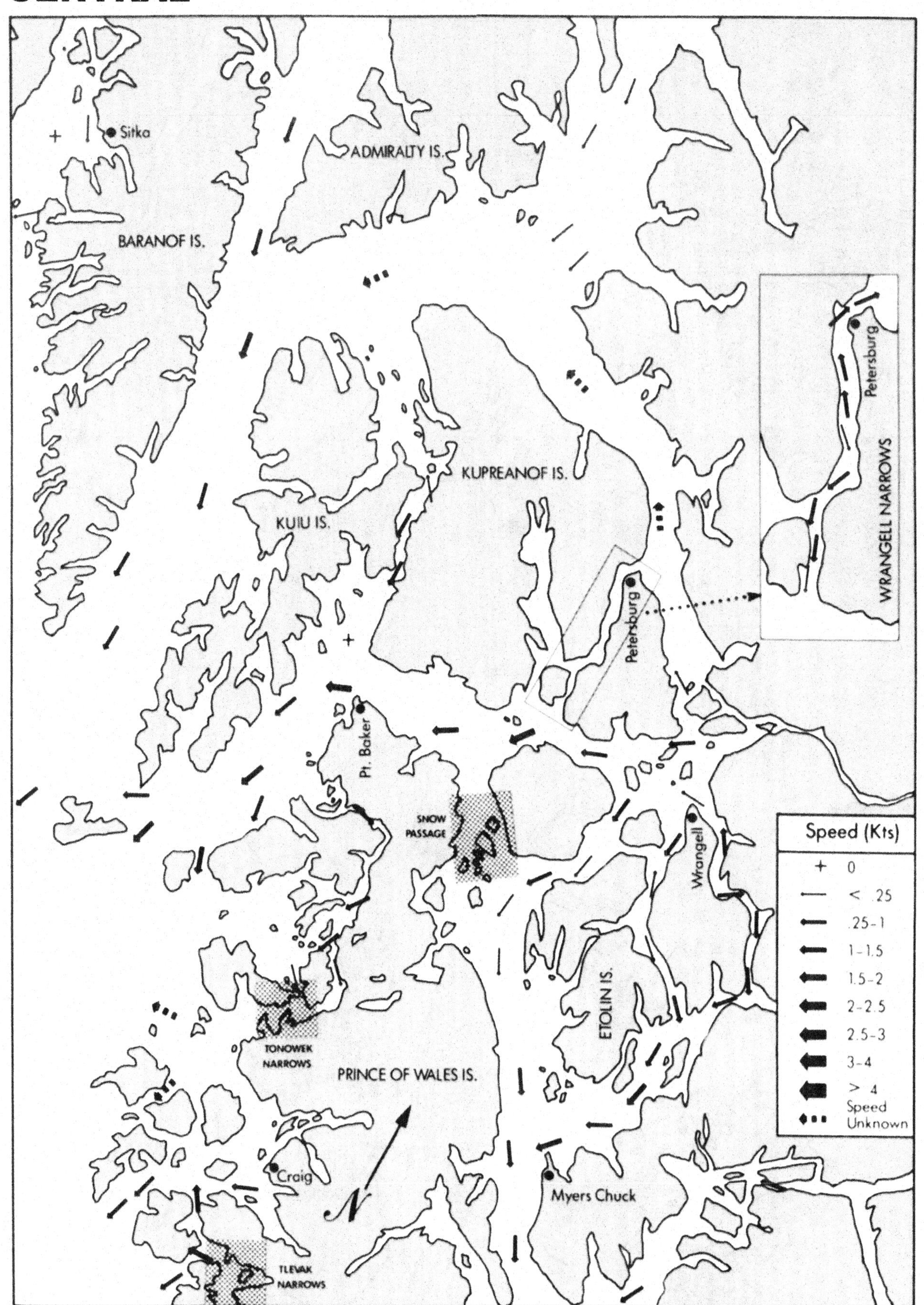

Sitka
ADMIRALTY IS.
BARANOF IS.
KUPREANOF IS.
KUIU IS.
Petersburg
WRANGELL NARROWS
Pt. Baker
SNOW PASSAGE
Wrangell
ETOLIN IS.
TONOWEK NARROWS
PRINCE OF WALES IS.
Craig
Myers Chuck
TLEVAK NARROWS
Speed (Kts)
0
< .25
.25-1
1-1.5
1.5-2
2-2.5
2.5-3
3-4
> 4
Speed Unknown

Shaded Areas: See SELECTED CHANNELS

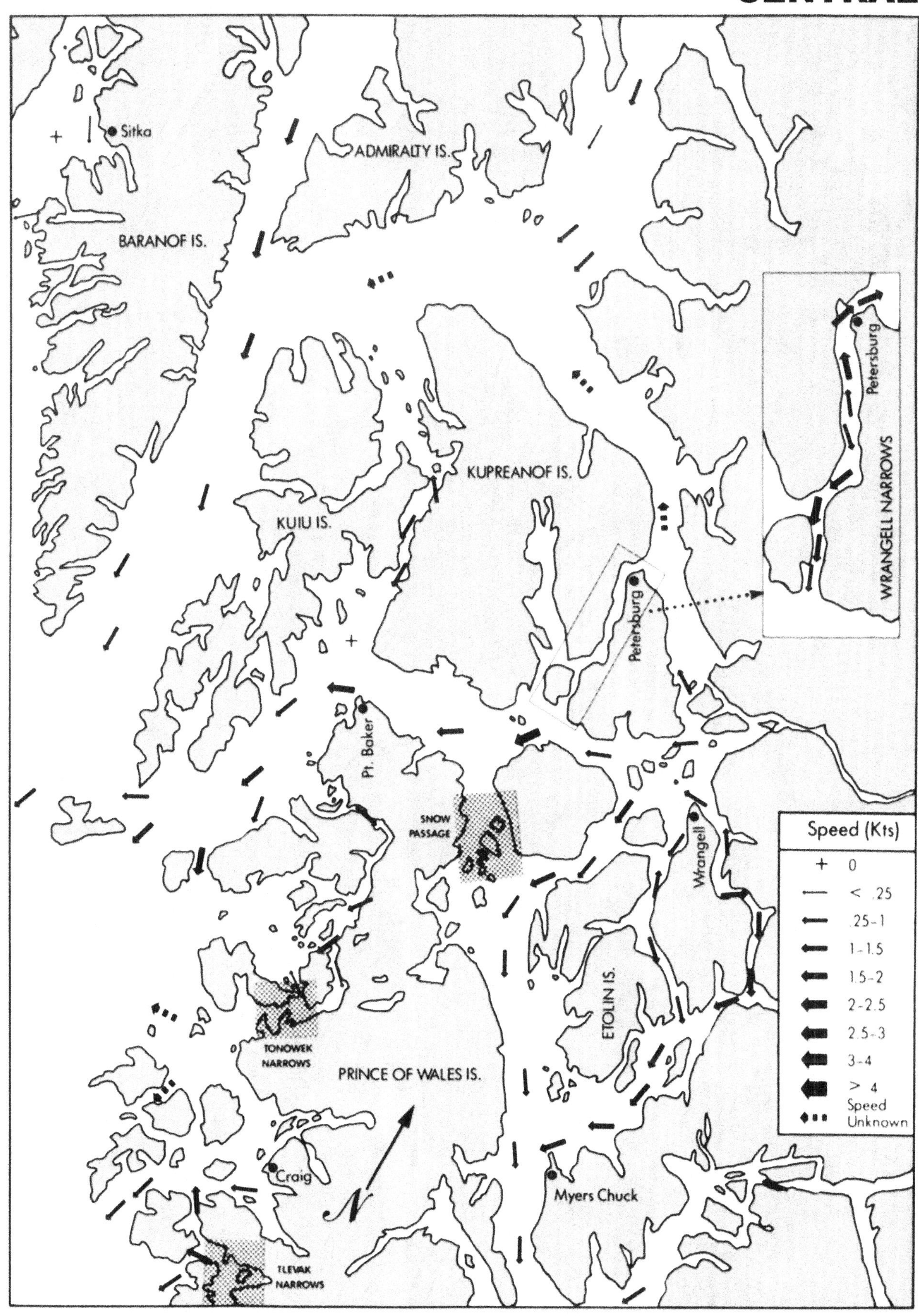
Sitka
ADMIRALTY IS.
BARANOF IS.
KUPREANOF IS.
KUIU IS.
Petersburg
Petersburg
WRANGELL NARROWS
Pt. Baker
SNOW PASSAGE
Wrangell
Speed (Kts)
+ 0
< .25
.25-1
1-1.5
1.5-2
2-2.5
2.5-3
3-4
> 4
Speed Unknown
ETOLIN IS.
TONOWEK NARROWS
PRINCE OF WALES IS.
Craig
Myers Chuck
TLEVAK NARROWS

Shaded Areas: See SELECTED CHANNELS

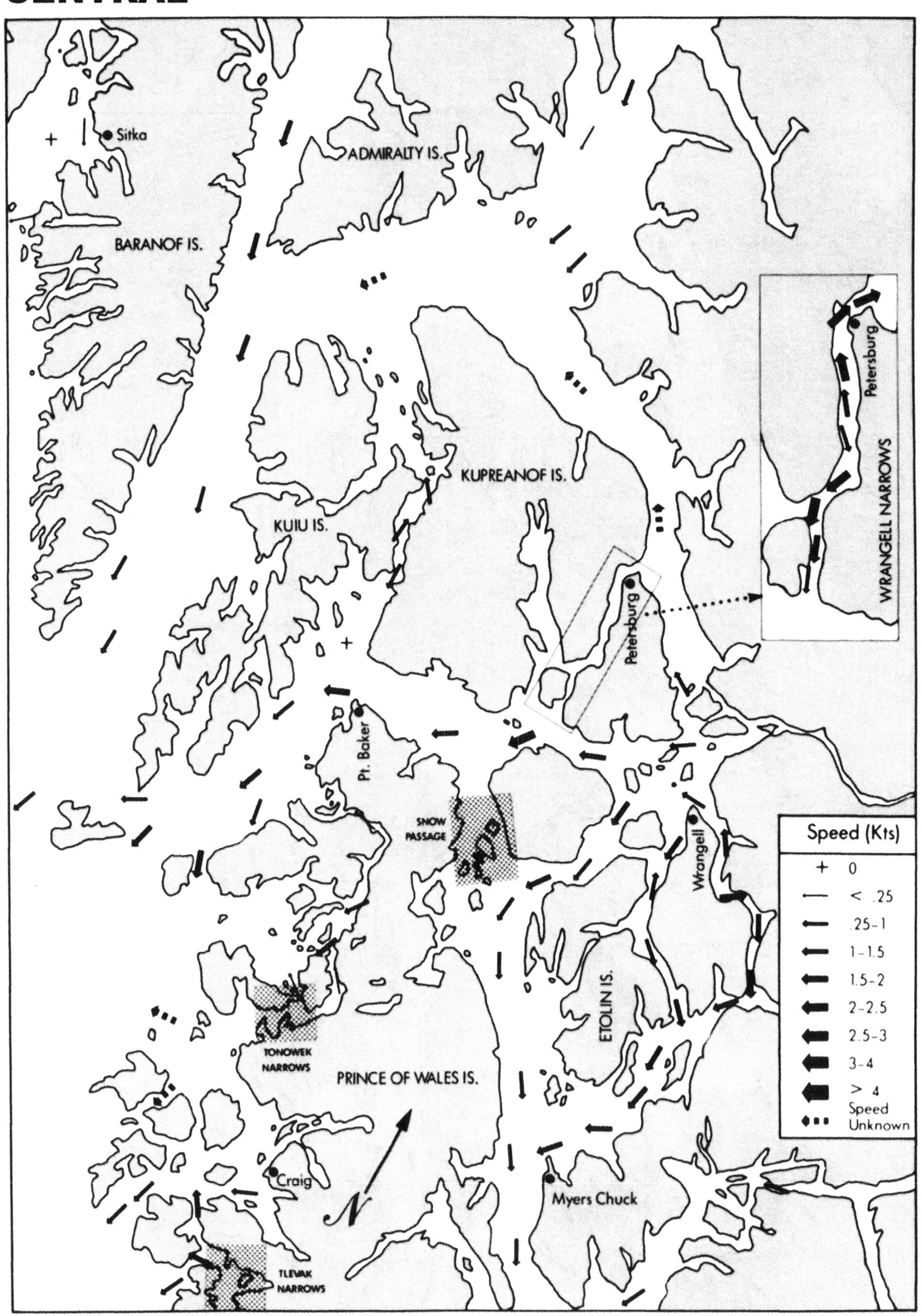
Sitka
ADMIRALTY IS.
BARANOF IS.
KUPREANOF IS.
KUIU IS.
Petersburg
WRANGELL NARROWS
Pt. Baker
SNOW PASSAGE
Wrangell
ETOLIN IS.
TONOWEK NARROWS
PRINCE OF WALES IS.
Craig
Myers Chuck
TLEVAK NARROWS
N
Speed (Kts)
0
< .25
.25-1
1-1.5
1.5-2
2-2.5
2.5-3
3-4
> 4
Speed Unknown

Shaded Areas: See SELECTED CHANNELS

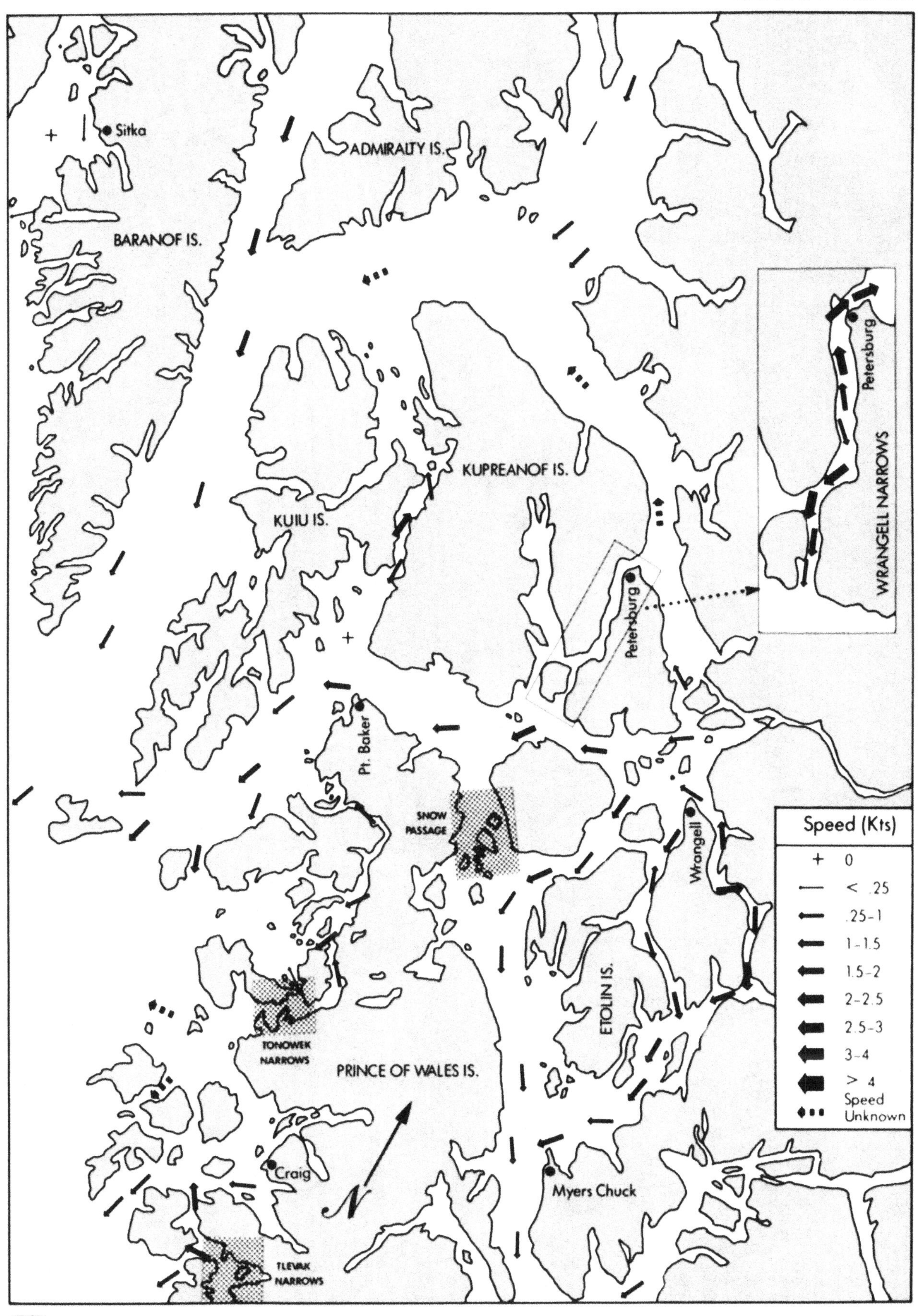
Sitka
ADMIRALTY IS.
BARANOF IS.
KUPREANOF IS.
KUIU IS.
Petersburg
WRANGELL NARROWS
Pt. Baker
SNOW PASSAGE
Wrangell
ETOLIN IS.
TONOWEK NARROWS
PRINCE OF WALES IS.
Craig
Myers Chuck
TLEVAK NARROWS
N
Speed (Kts)
0
< .25
.25-1
1-1.5
1.5-2
2-2.5
2.5-3
3-4
> 4
Speed Unknown

Shaded Areas: See SELECTED CHANNELS

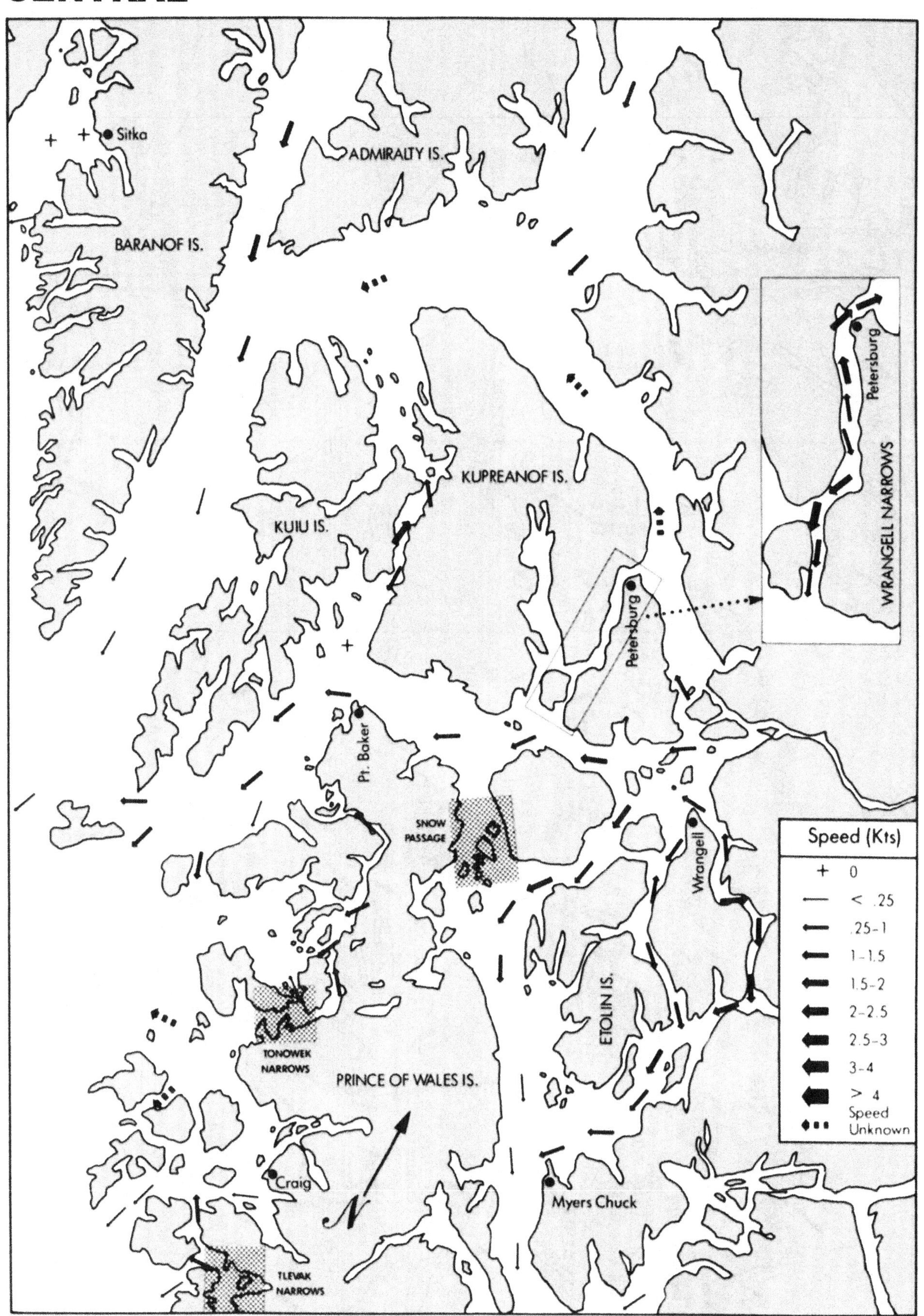
Sitka
ADMIRALTY IS.
BARANOF IS.
KUPREANOF IS.
KUIU IS.
Petersburg
Petersburg
WRANGELL NARROWS
Pt. Baker
SNOW PASSAGE
Wrangell
Speed (Kts)
+ 0
< .25
.25-1
1-1.5
1.5-2
2-2.5
2.5-3
3-4
> 4
Speed Unknown
ETOLIN IS.
TONOWEK NARROWS
PRINCE OF WALES IS.
Craig
Myers Chuck
TLEVAK NARROWS
N

Shaded Areas: See SELECTED CHANNELS

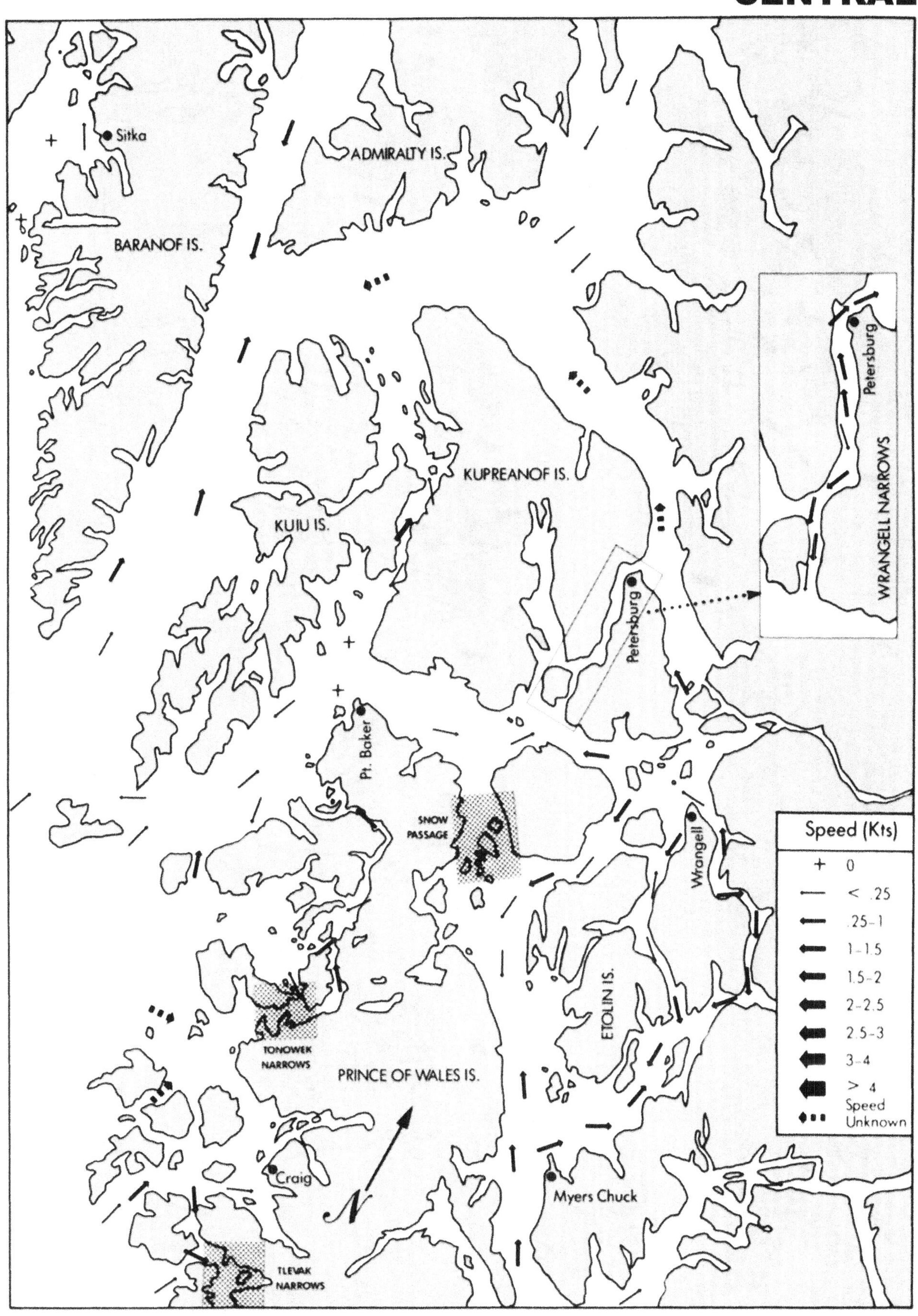
Sitka
ADMIRALTY IS.
BARANOF IS.
KUPREANOF IS.
KUIU IS.
Petersburg
Pt. Baker
SNOW PASSAGE
Wrangell
ETOLIN IS.
TONOWEK NARROWS
PRINCE OF WALES IS.
Craig
Myers Chuck
TLEVAK NARROWS
N
WRANGELL NARROWS
Speed (Kts)
0
< .25
.25-1
1-1.5
1.5-2
2-2.5
2.5-3
3-4
> 4
Speed Unknown

Shaded Areas: See SELECTED CHANNELS

SOUTH

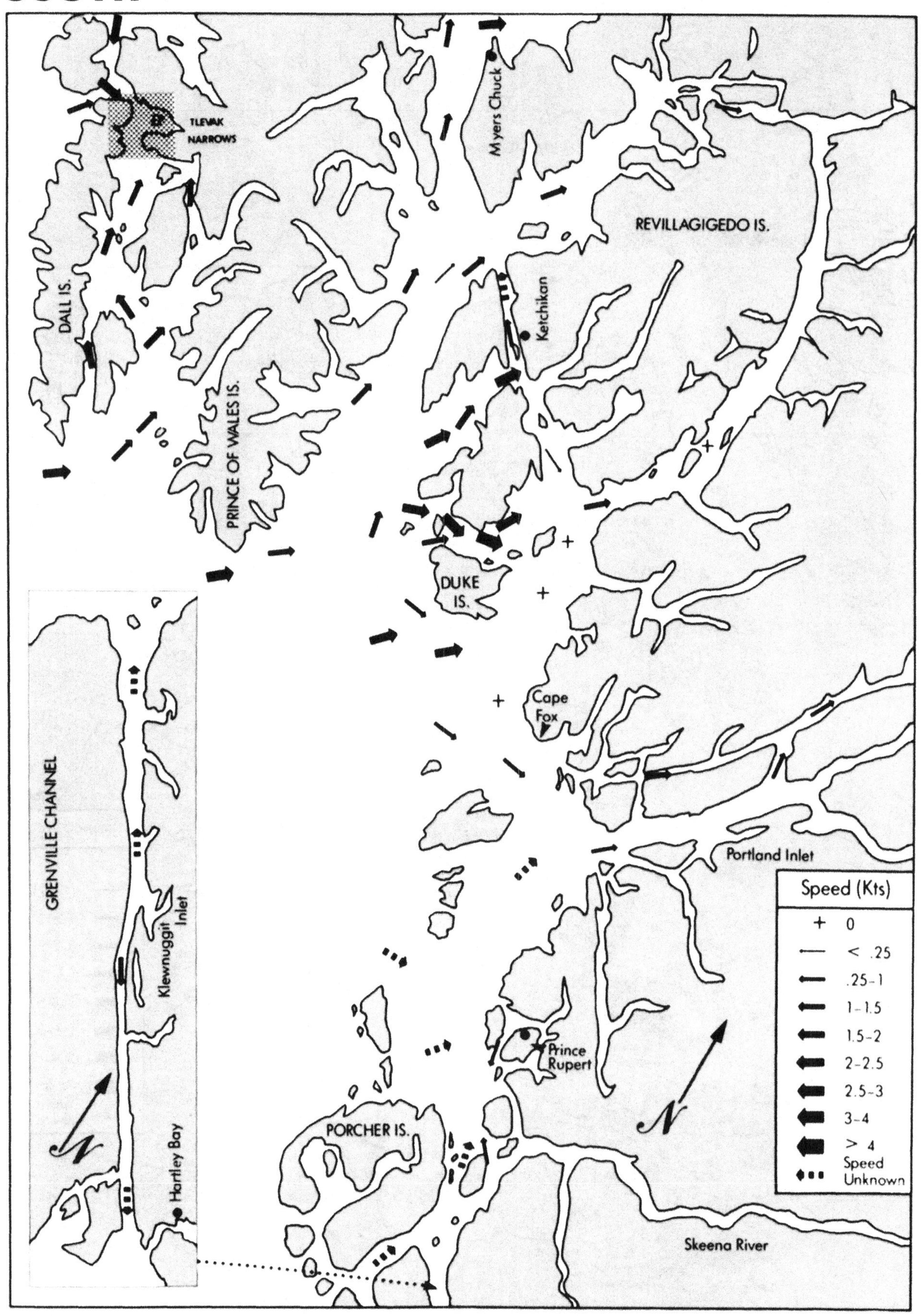

TLEVAK NARROWS
Myers Chuck
REVILLAGIGEDO IS.
DALL IS.
Ketchikan
PRINCE OF WALES IS.
DUKE IS.
Cape Fox
GRENVILLE CHANNEL
Klewnuggit Inlet
Portland Inlet
Speed (Kts)
0
< .25
.25-1
1-1.5
1.5-2
2-2.5
2.5-3
3-4
> 4
Speed Unknown
Prince Rupert
Hartley Bay
PORCHER IS.
Skeena River

Shaded Areas: See SELECTED CHANNELS

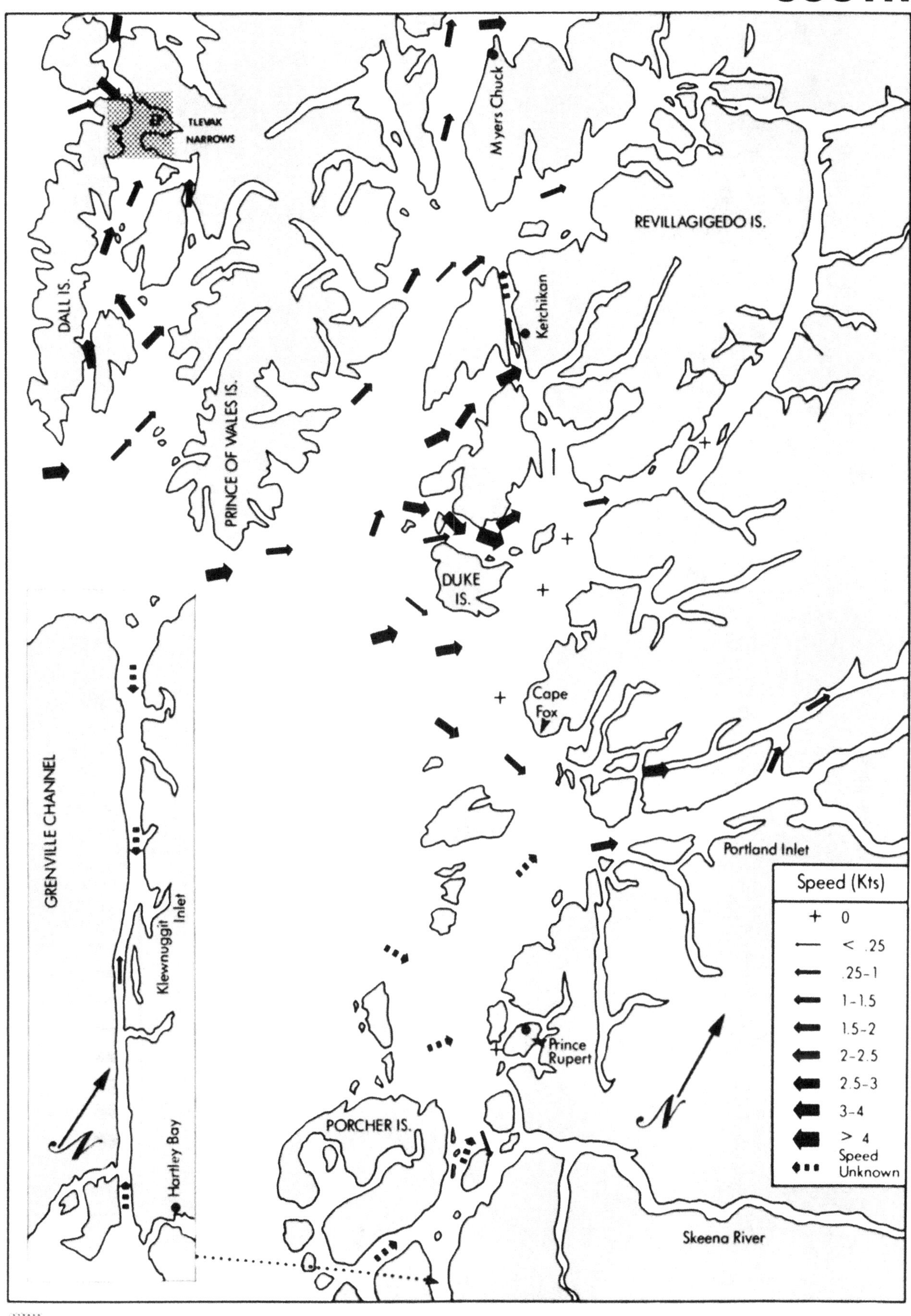

TLEVAK NARROWS
Myers Chuck
REVILLAGIGEDO IS.
DALL IS.
Ketchikan
PRINCE OF WALES IS.
DUKE IS.
Cape Fox
GRENVILLE CHANNEL
Portland Inlet
Speed (Kts)
+ 0
< .25
.25-1
1-1.5
1.5-2
2-2.5
2.5-3
3-4
> 4
Speed Unknown
Klewnuggit Inlet
Prince Rupert
PORCHER IS.
Hartley Bay
Skeena River

Shaded Areas: See SELECTED CHANNELS

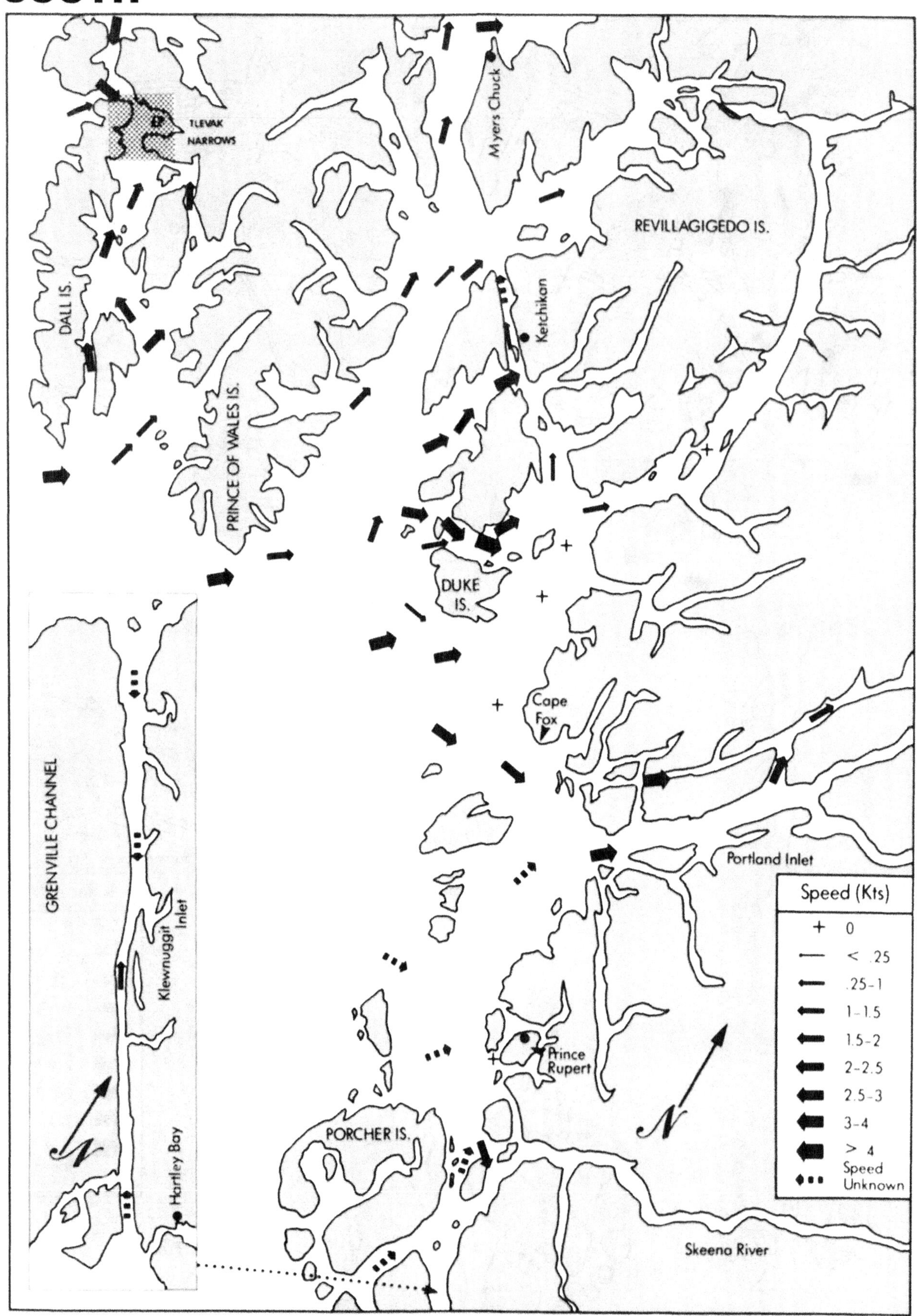
TLEVAK NARROWS
Myers Chuck
REVILLAGIGEDO IS.
DALL IS.
Ketchikan
PRINCE OF WALES IS.
DUKE IS.
Cape Fox
GRENVILLE CHANNEL
Klewnuggit Inlet
Portland Inlet
Speed (Kts)
0
< .25
.25-1
1-1.5
1.5-2
2-2.5
2.5-3
3-4
> 4
Speed Unknown
Prince Rupert
PORCHER IS.
Hartley Bay
Skeena River

Shaded Areas: See SELECTED CHANNELS

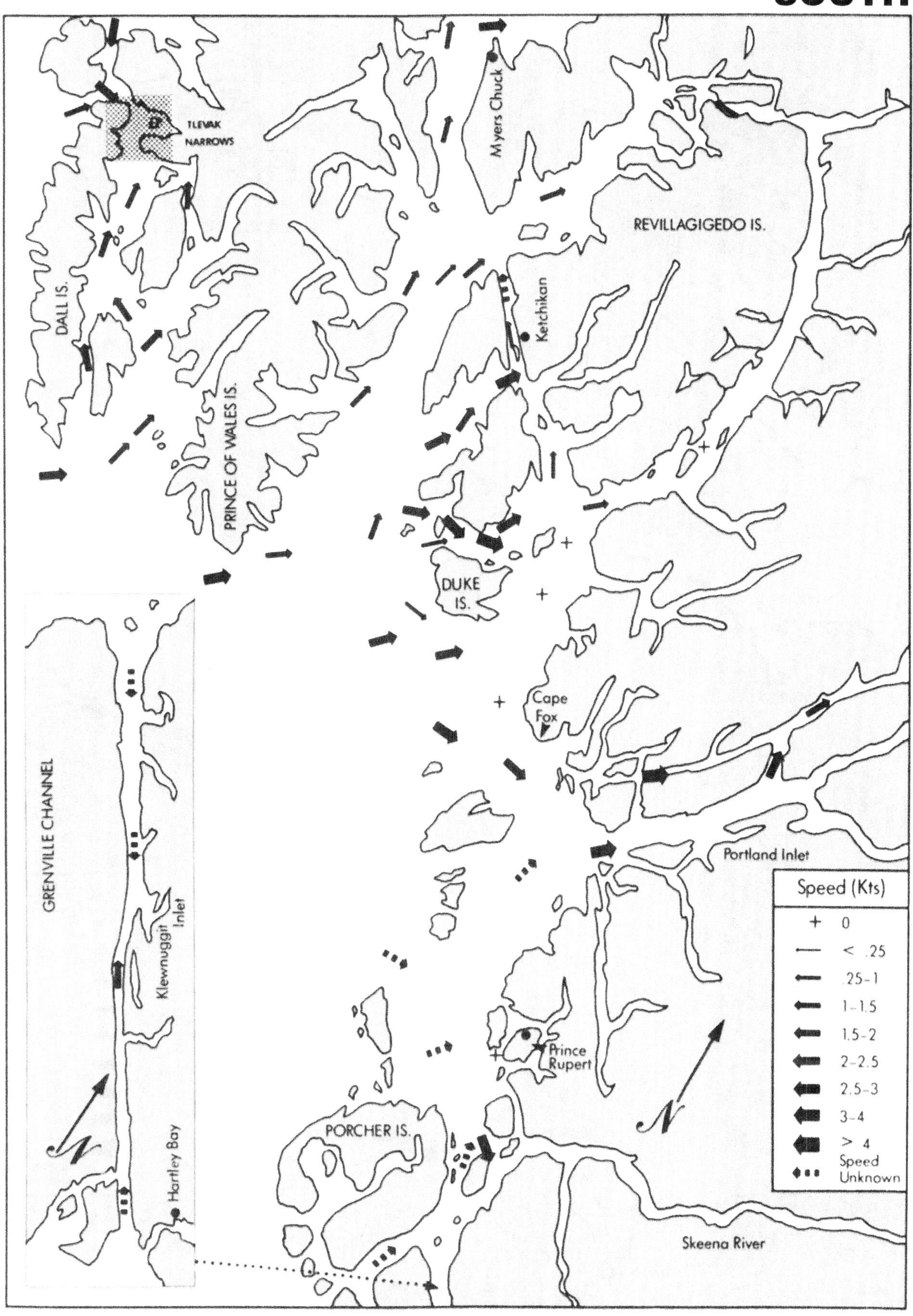
TLEVAK NARROWS
Myers Chuck
REVILLAGIGEDO IS.
DALL IS.
Ketchikan
PRINCE OF WALES IS.
DUKE IS.
Cape Fox
GRENVILLE CHANNEL
Klewnuggit Inlet
Portland Inlet
Speed (Kts)
0
< .25
.25-1
1-1.5
1.5-2
2-2.5
2.5-3
3-4
> 4
Speed Unknown
Prince Rupert
PORCHER IS.
Hartley Bay
Skeena River

Shaded Areas: See SELECTED CHANNELS

SOUTH

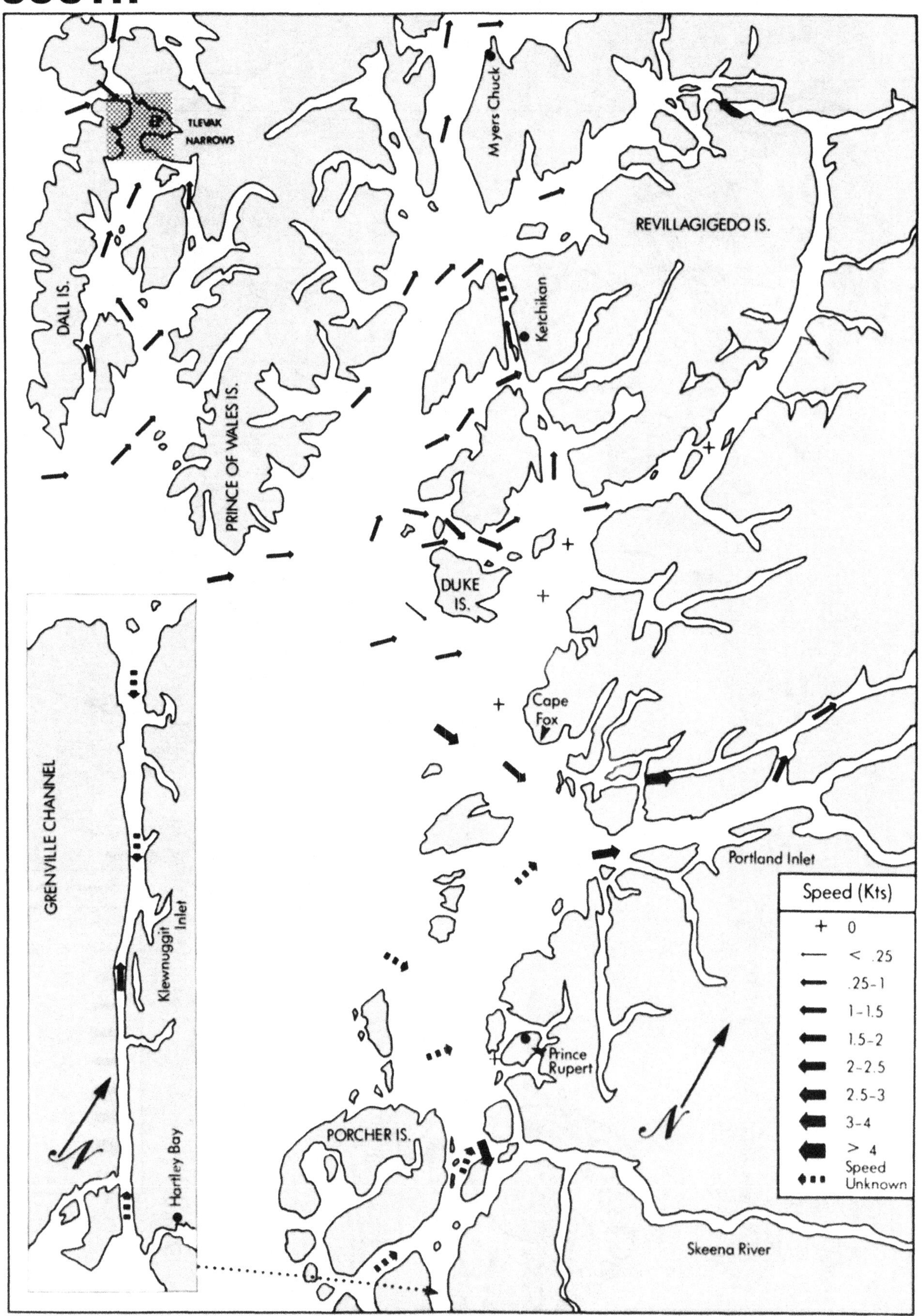
TLEVAK NARROWS
Myers Chuck
REVILLAGIGEDO IS.
DALL IS.
Ketchikan
PRINCE OF WALES IS.
DUKE IS.
Cape Fox
GRENVILLE CHANNEL
Klewnuggit Inlet
Hartley Bay
Portland Inlet
Speed (Kts)
+ 0
< .25
.25-1
1-1.5
1.5-2
2-2.5
2.5-3
3-4
> 4
Speed Unknown
Prince Rupert
PORCHER IS.
Skeena River

Shaded Areas: See SELECTED CHANNELS

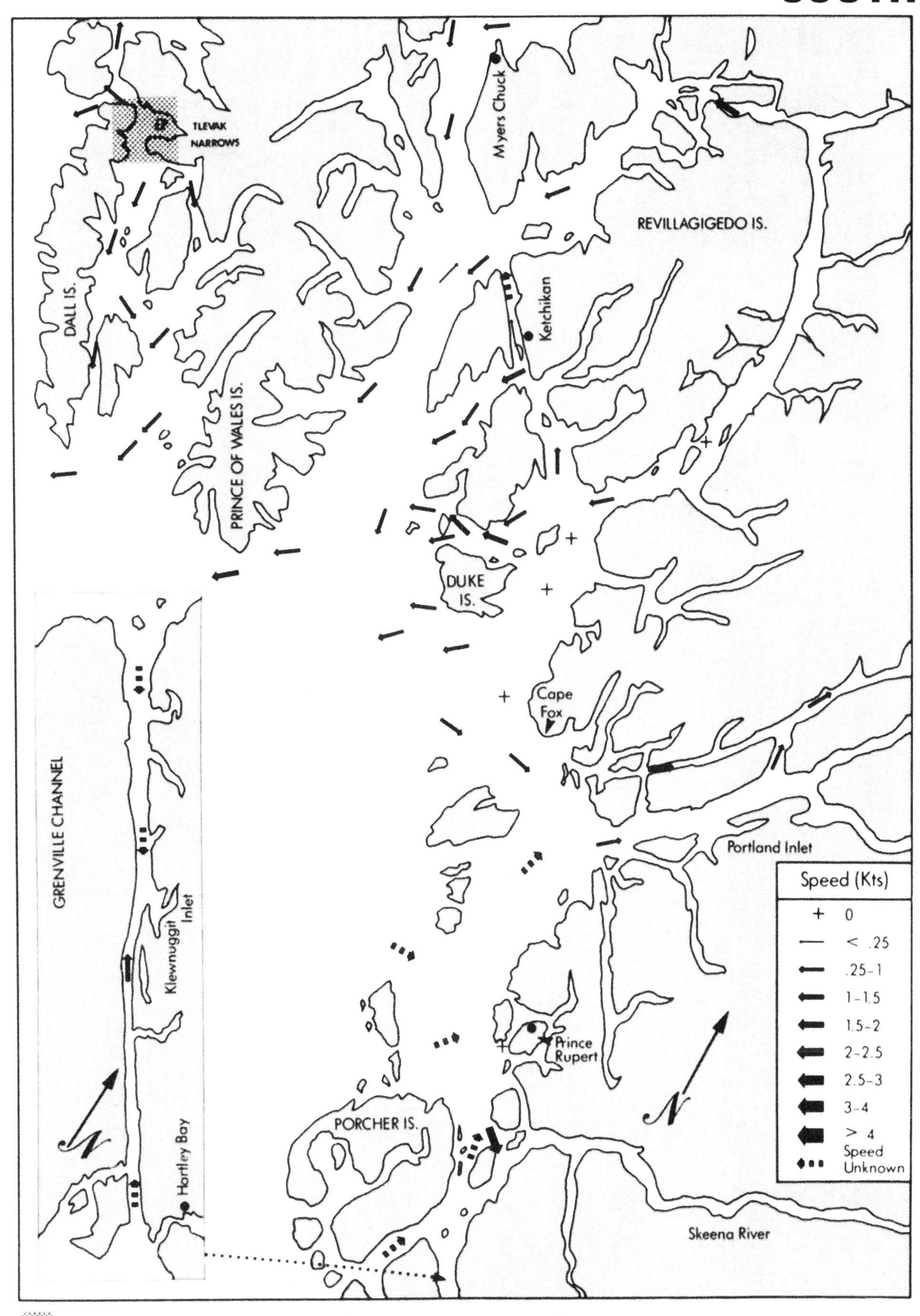
TLEVAK NARROWS
Myers Chuck
REVILLAGIGEDO IS.
DALL IS.
Ketchikan
PRINCE OF WALES IS.
DUKE IS.
Cape Fox
GRENVILLE CHANNEL
Klewnuggit Inlet
Hartley Bay
Portland Inlet
Prince Rupert
PORCHER IS.
Skeena River
Speed (Kts)
0
< .25
.25-1
1-1.5
1.5-2
2-2.5
2.5-3
3-4
> 4
Speed Unknown

Shaded Areas: See SELECTED CHANNELS

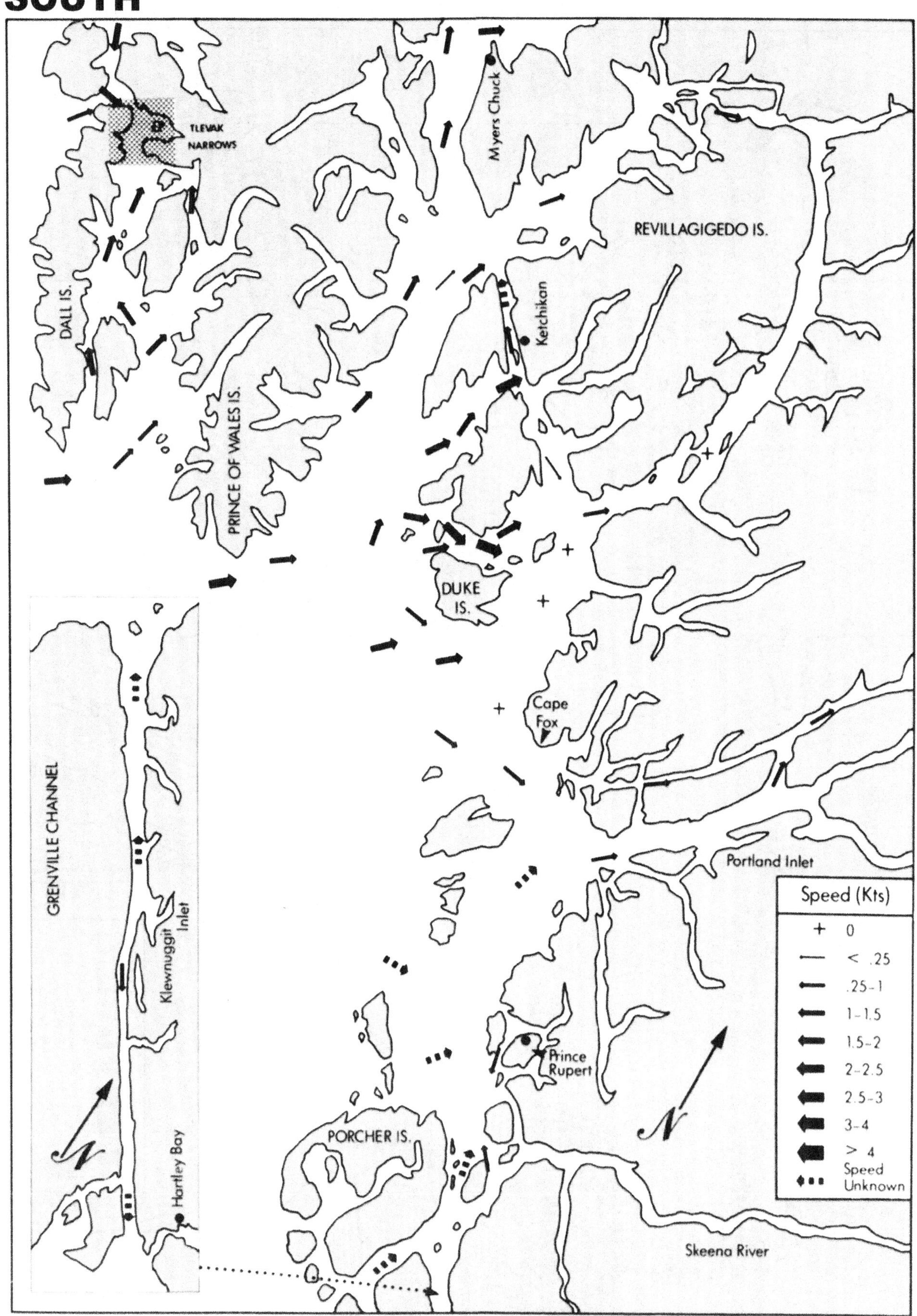
TLEVAK NARROWS
DALL IS.
PRINCE OF WALES IS.
Myers Chuck
Ketchikan
REVILLAGIGEDO IS.
DUKE IS.
Cape Fox
Portland Inlet
Prince Rupert
PORCHER IS.
Skeena River
GRENVILLE CHANNEL
Klewnuggit Inlet
Hartley Bay
Speed (Kts)
0
< .25
.25–1
1–1.5
1.5–2
2–2.5
2.5–3
3–4
> 4
Speed Unknown

Shaded Areas: See SELECTED CHANNELS

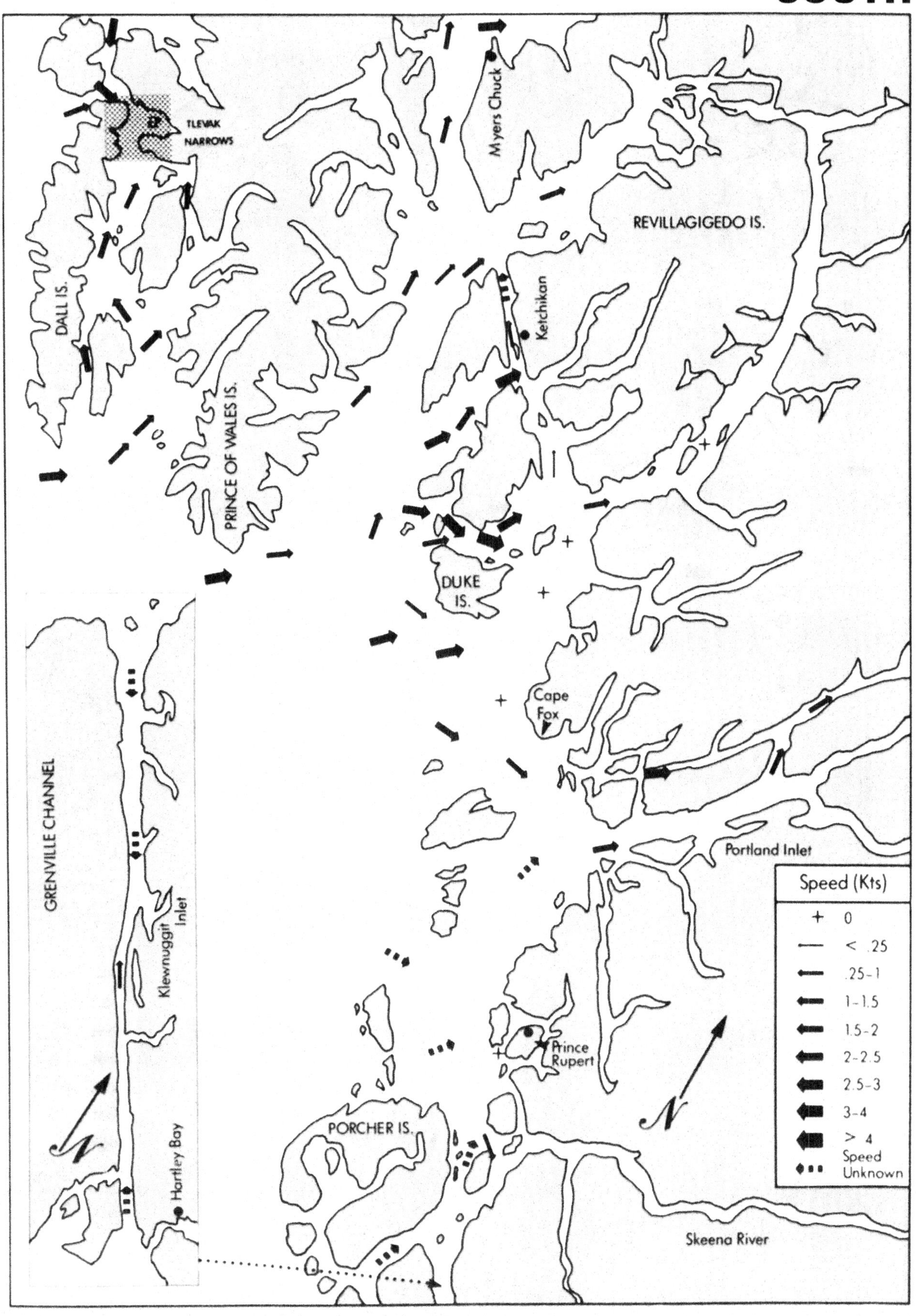
TLEVAK NARROWS
Myers Chuck
REVILLAGIGEDO IS.
DALL IS.
Ketchikan
PRINCE OF WALES IS.
DUKE IS.
Cape Fox
GRENVILLE CHANNEL
Klewnuggit Inlet
Portland Inlet
Speed (Kts)
0
< .25
.25-1
1-1.5
1.5-2
2-2.5
2.5-3
3-4
> 4
Speed Unknown
Prince Rupert
PORCHER IS.
Hartley Bay
Skeena River

Shaded Areas: See SELECTED CHANNELS

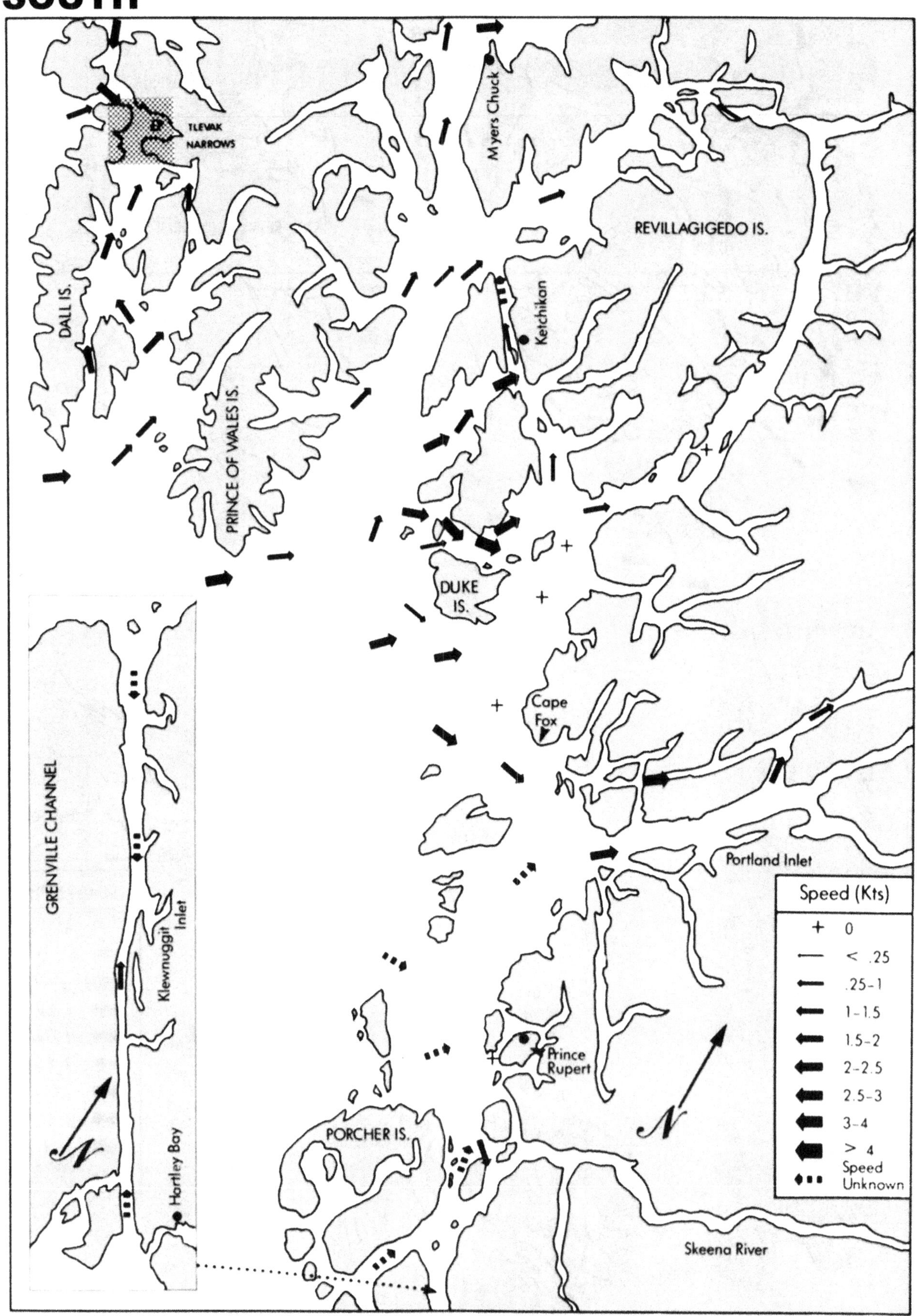
TLEVAK
NARROWS
Myers Chuck
REVILLAGIGEDO IS.
DALL IS.
Ketchikan
PRINCE OF WALES IS.
DUKE
IS.
Cape
Fox
GRENVILLE CHANNEL
Klewnuggit
Inlet
Portland Inlet
Speed (Kts)
0
< .25
.25-1
1-1.5
1.5-2
2-2.5
2.5-3
3-4
> 4
Speed
Unknown
Prince
Rupert
PORCHER IS.
Hartley Bay
Skeena River

Shaded Areas: See SELECTED CHANNELS

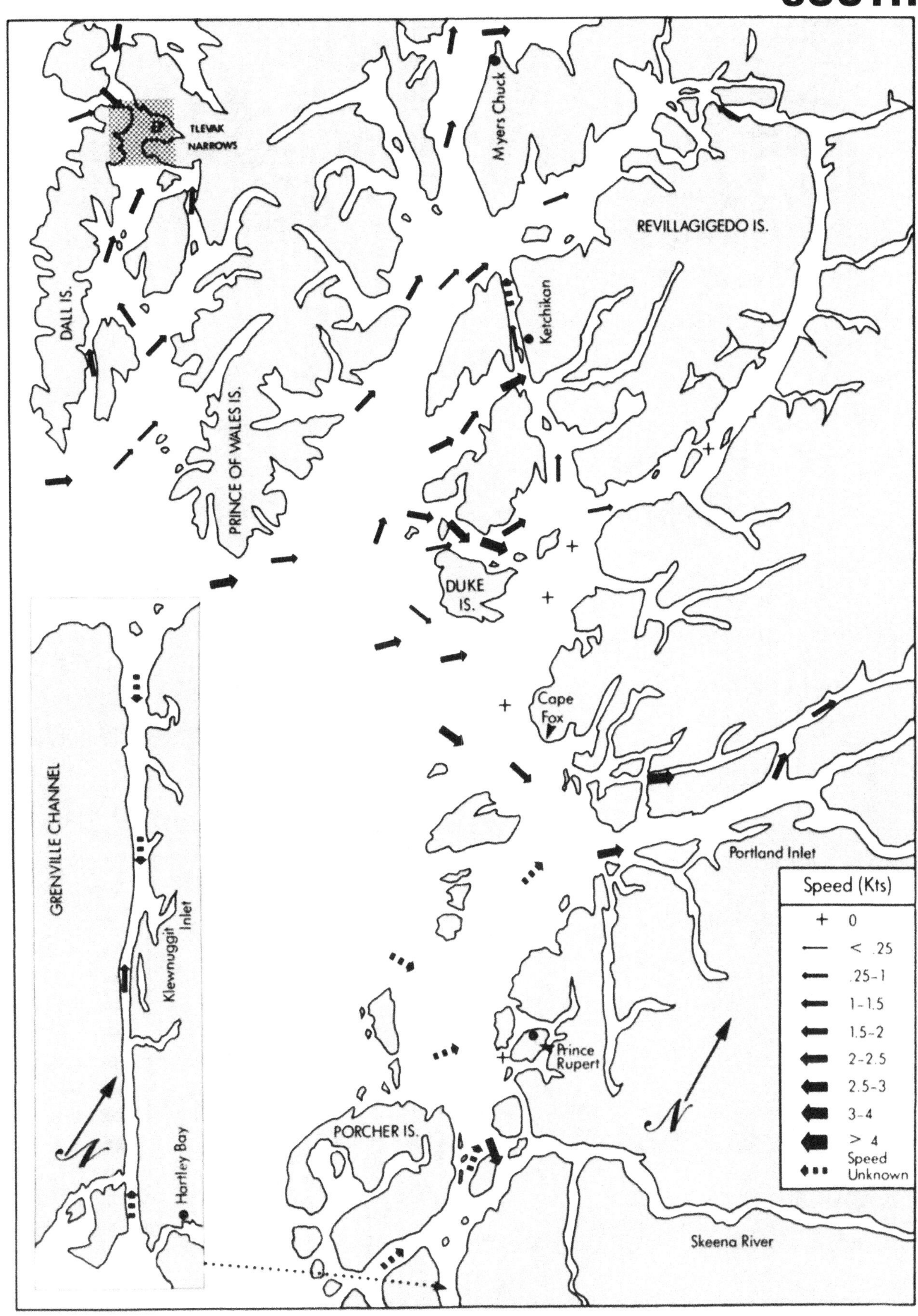
TLEVAK NARROWS
Myers Chuck
REVILLAGIGEDO IS.
DALL IS.
Ketchikan
PRINCE OF WALES IS.
DUKE IS.
Cape Fox
GRENVILLE CHANNEL
Klewnuggit Inlet
Portland Inlet
Speed (Kts)
0
< .25
.25-1
1-1.5
1.5-2
2-2.5
2.5-3
3-4
> 4
Speed Unknown
Prince Rupert
PORCHER IS.
Hartley Bay
Skeena River

Shaded Areas: See SELECTED CHANNELS

SOUTH

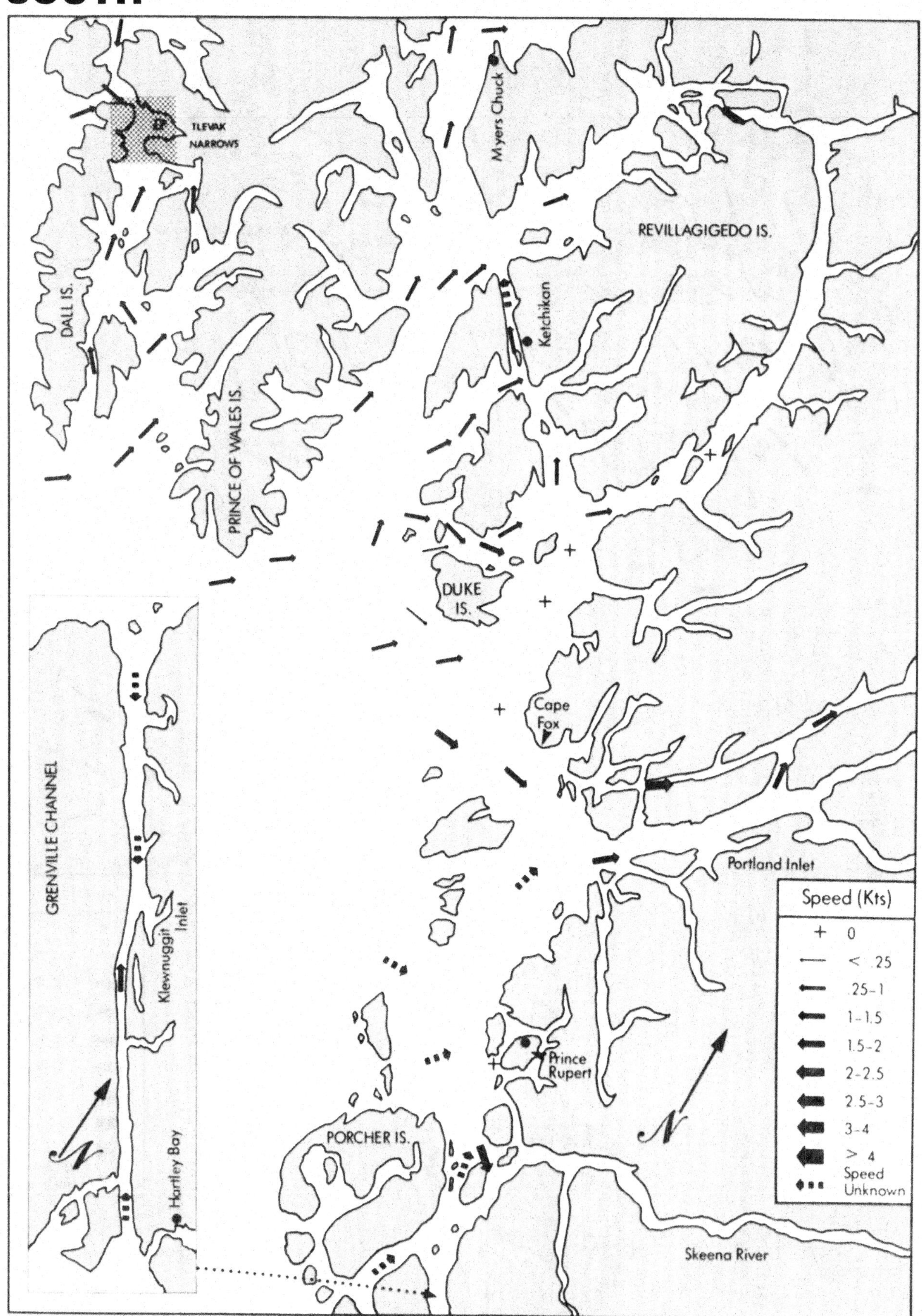

TLEVAK
NARROWS
DALL IS.
PRINCE OF WALES IS.
Myers Chuck
REVILLAGIGEDO IS.
Ketchikan
DUKE IS.
Cape Fox
Portland Inlet
GRENVILLE CHANNEL
Klewnuggit Inlet
Hartley Bay
Prince Rupert
PORCHER IS.
Skeena River
Speed (Kts)
0
< .25
.25-1
1-1.5
1.5-2
2-2.5
2.5-3
3-4
> 4
Speed Unknown

Shaded Areas: See SELECTED CHANNELS

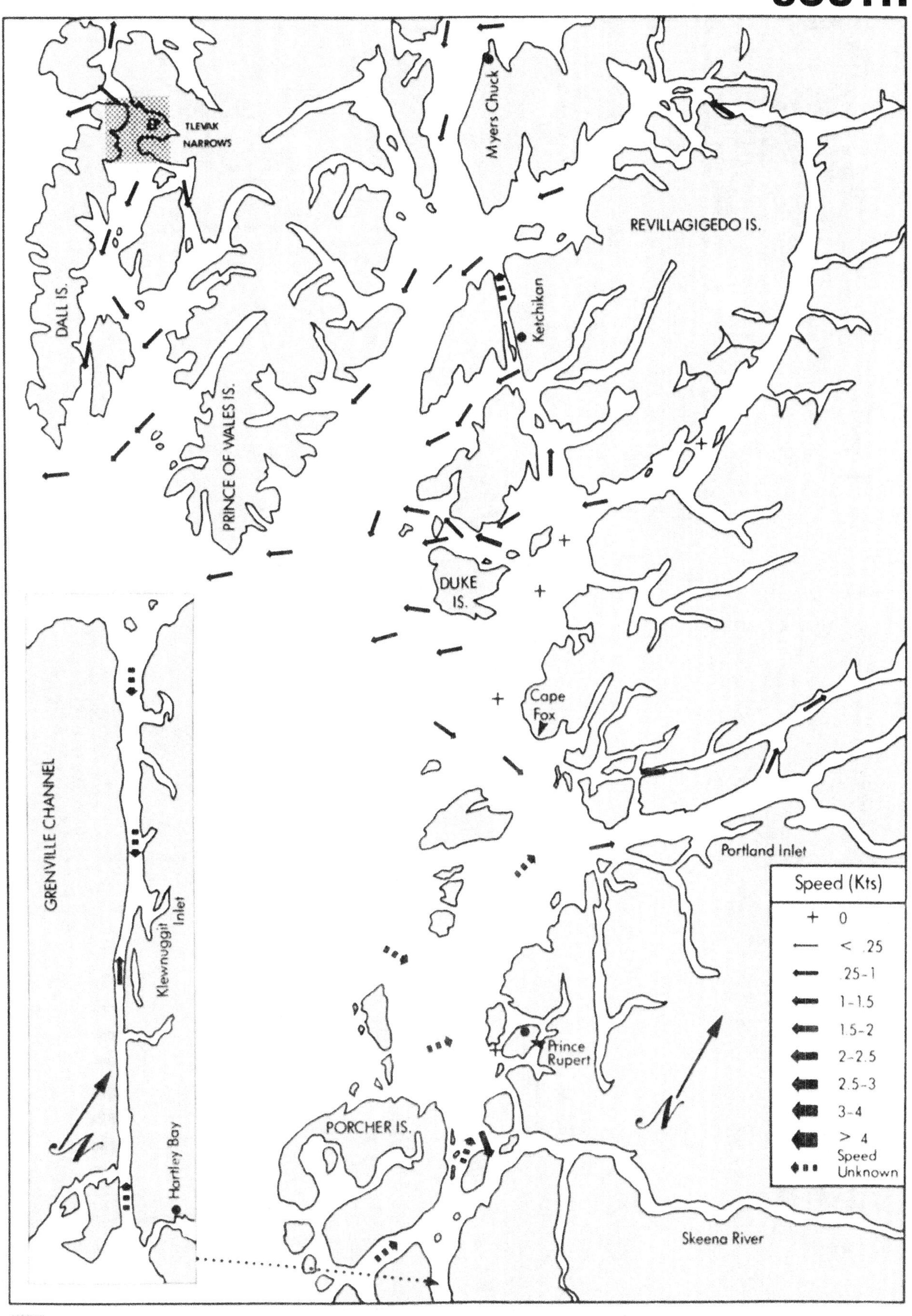
TLEVAK
NARROWS
Myers Chuck
REVILLAGIGEDO IS.
DALL IS.
Ketchikan
PRINCE OF WALES IS.
DUKE
IS.
Cape
Fox
GRENVILLE CHANNEL
Klewnuggit
Inlet
Portland Inlet
Speed (Kts)
0
< .25
.25-1
1-1.5
1.5-2
2-2.5
2.5-3
3-4
> 4
Speed
Unknown
Prince
Rupert
PORCHER IS.
Hartley Bay
Skeena River

Shaded Areas: See SELECTED CHANNELS

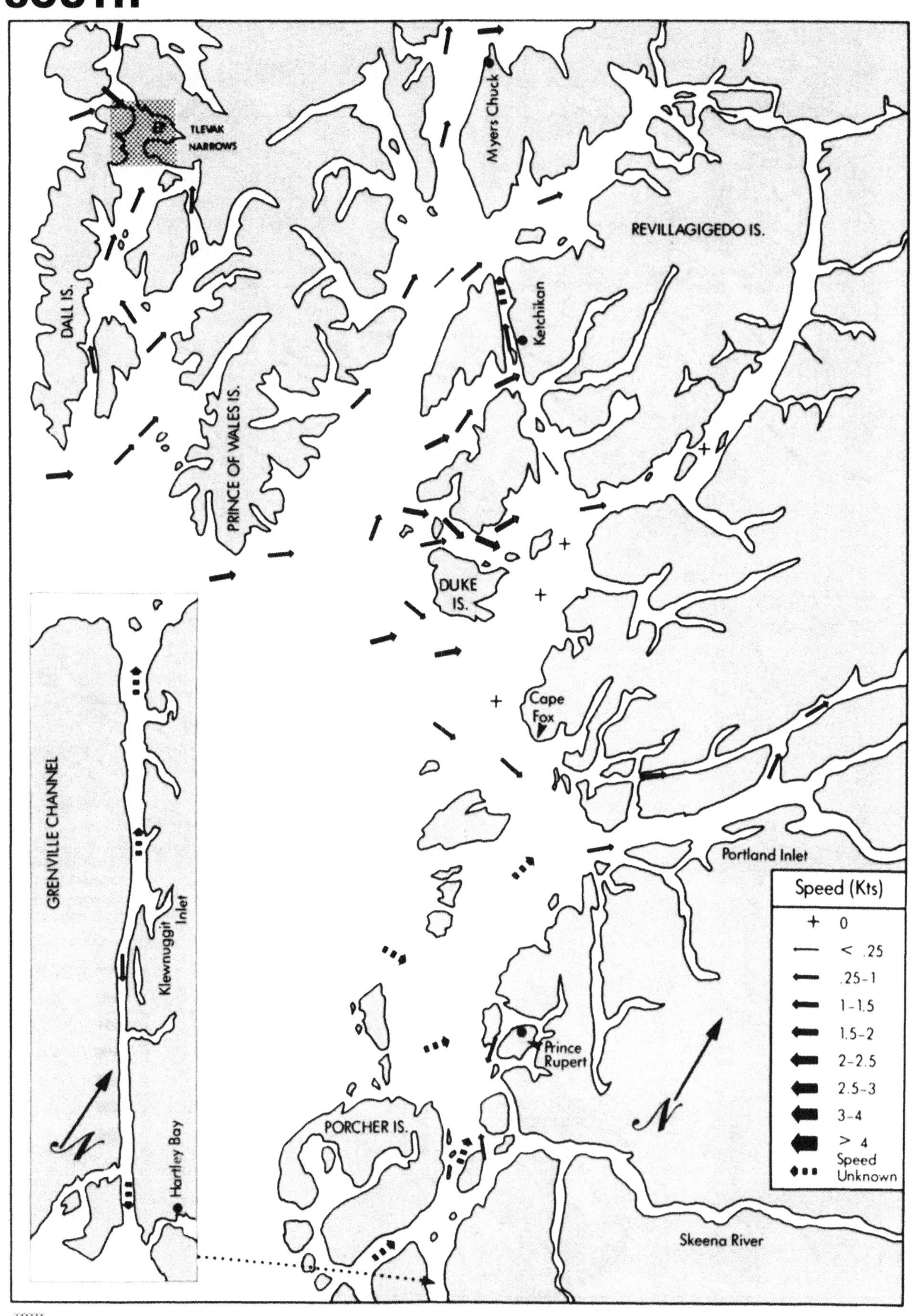
TLEVAK
NARROWS
Myers Chuck
REVILLAGIGEDO IS.
DALL IS.
Ketchikan
PRINCE OF WALES IS.
DUKE IS.
Cape Fox
GRENVILLE CHANNEL
Klewnuggit Inlet
Portland Inlet
Speed (Kts)
0
< .25
.25-1
1-1.5
1.5-2
2-2.5
2.5-3
3-4
> 4
Speed Unknown
Prince Rupert
Hartley Bay
PORCHER IS.
Skeena River

Shaded Areas: See SELECTED CHANNELS

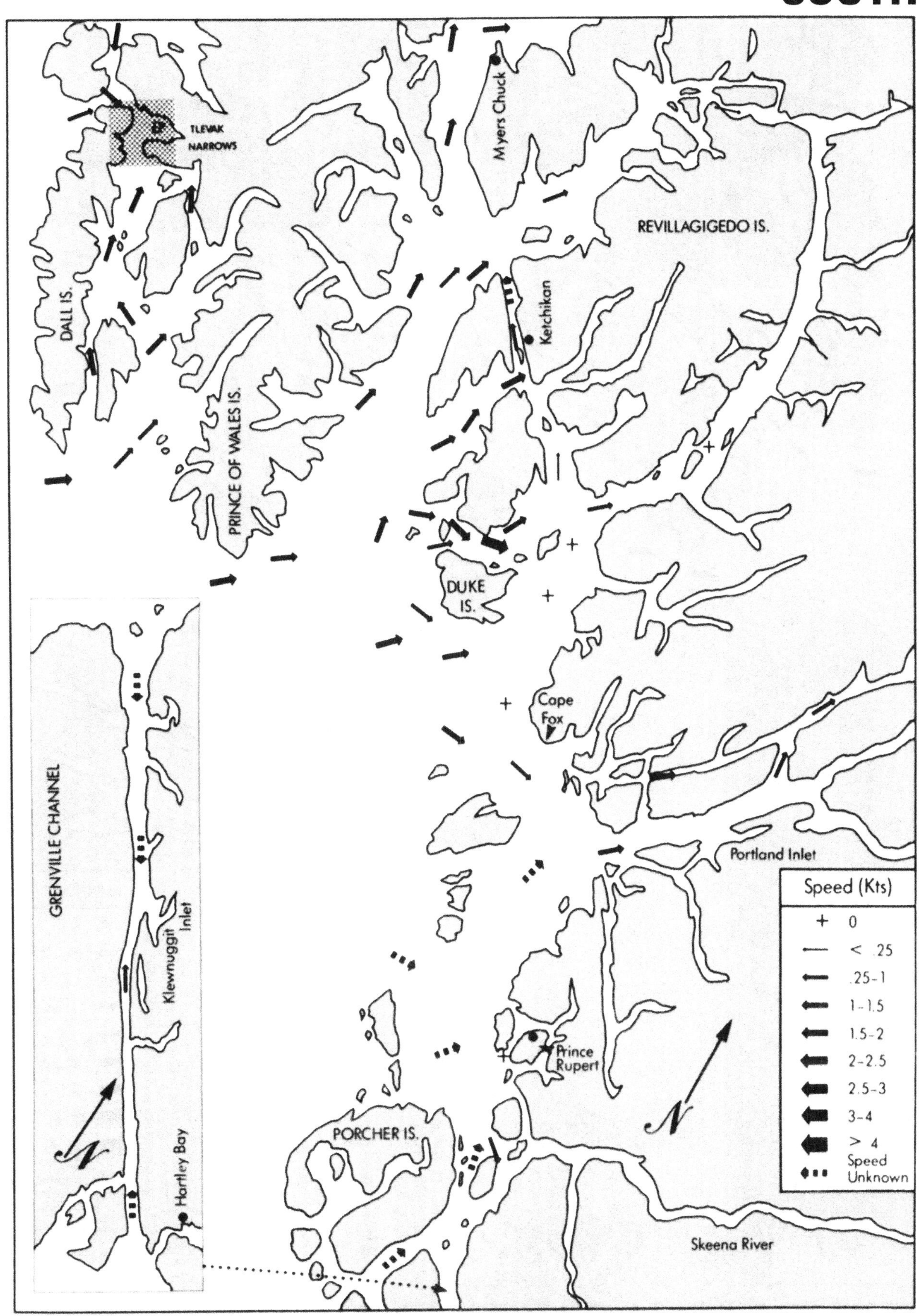
TLEVAK
NARROWS
Myers Chuck
REVILLAGIGEDO IS.
DALL IS.
Ketchikan
PRINCE OF WALES IS.
DUKE
IS.
Cape
Fox
GRENVILLE CHANNEL
Klewnuggit
Inlet
Portland Inlet
Speed (Kts)
0
< .25
.25-1
1-1.5
1.5-2
2-2.5
2.5-3
3-4
> 4
Speed
Unknown
Prince
Rupert
PORCHER IS.
Hartley Bay
Skeena River

Shaded Areas: See SELECTED CHANNELS

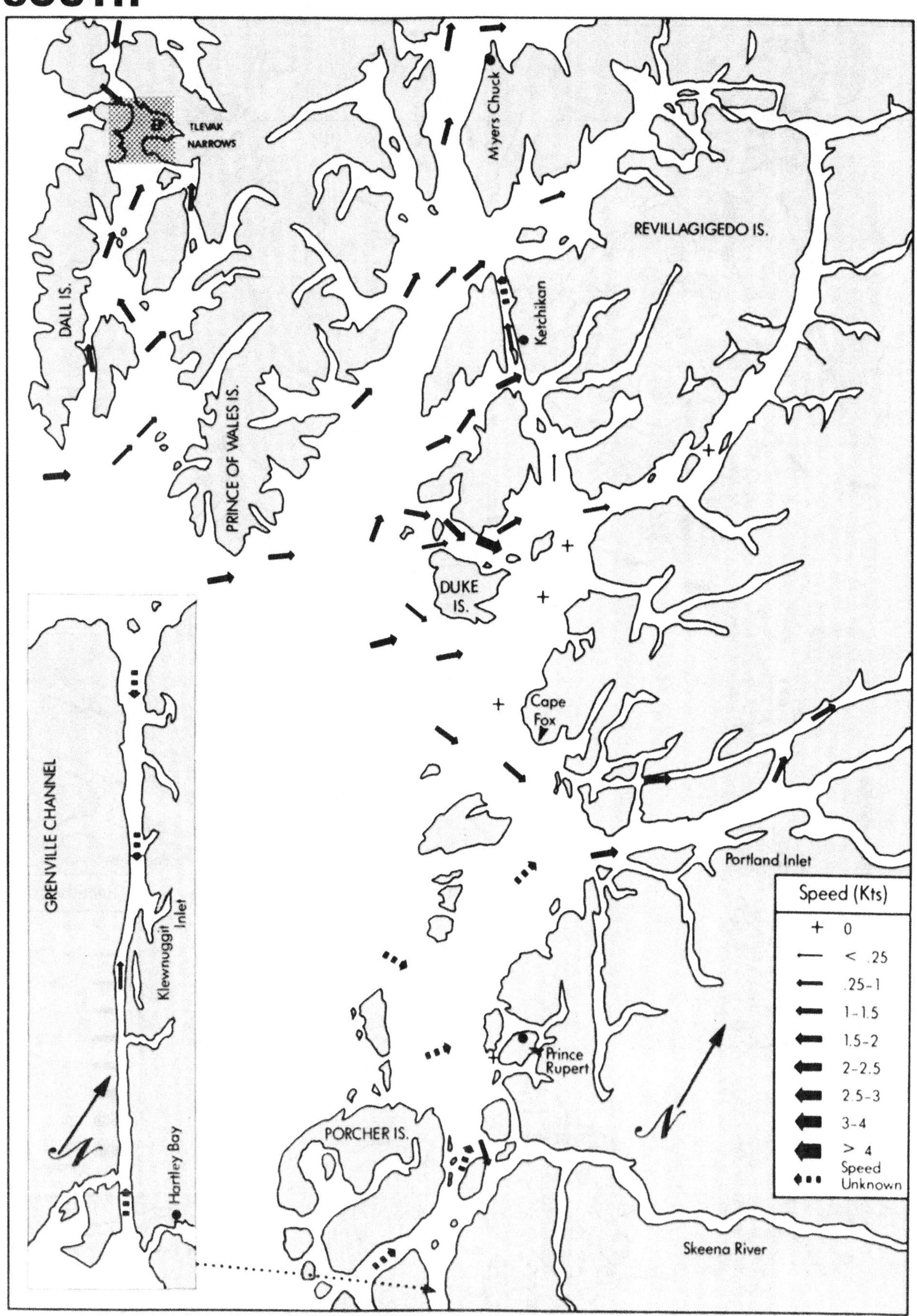
TLEVAK NARROWS
Myers Chuck
REVILLAGIGEDO IS.
DALL IS.
Ketchikan
PRINCE OF WALES IS.
DUKE IS.
Cape Fox
Portland Inlet
GRENVILLE CHANNEL
Klewnuggit Inlet
Hartley Bay
Prince Rupert
PORCHER IS.
Skeena River
Speed (Kts)
0
< .25
.25-1
1-1.5
1.5-2
2-2.5
2.5-3
3-4
> 4
Speed Unknown

Shaded Areas: See SELECTED CHANNELS

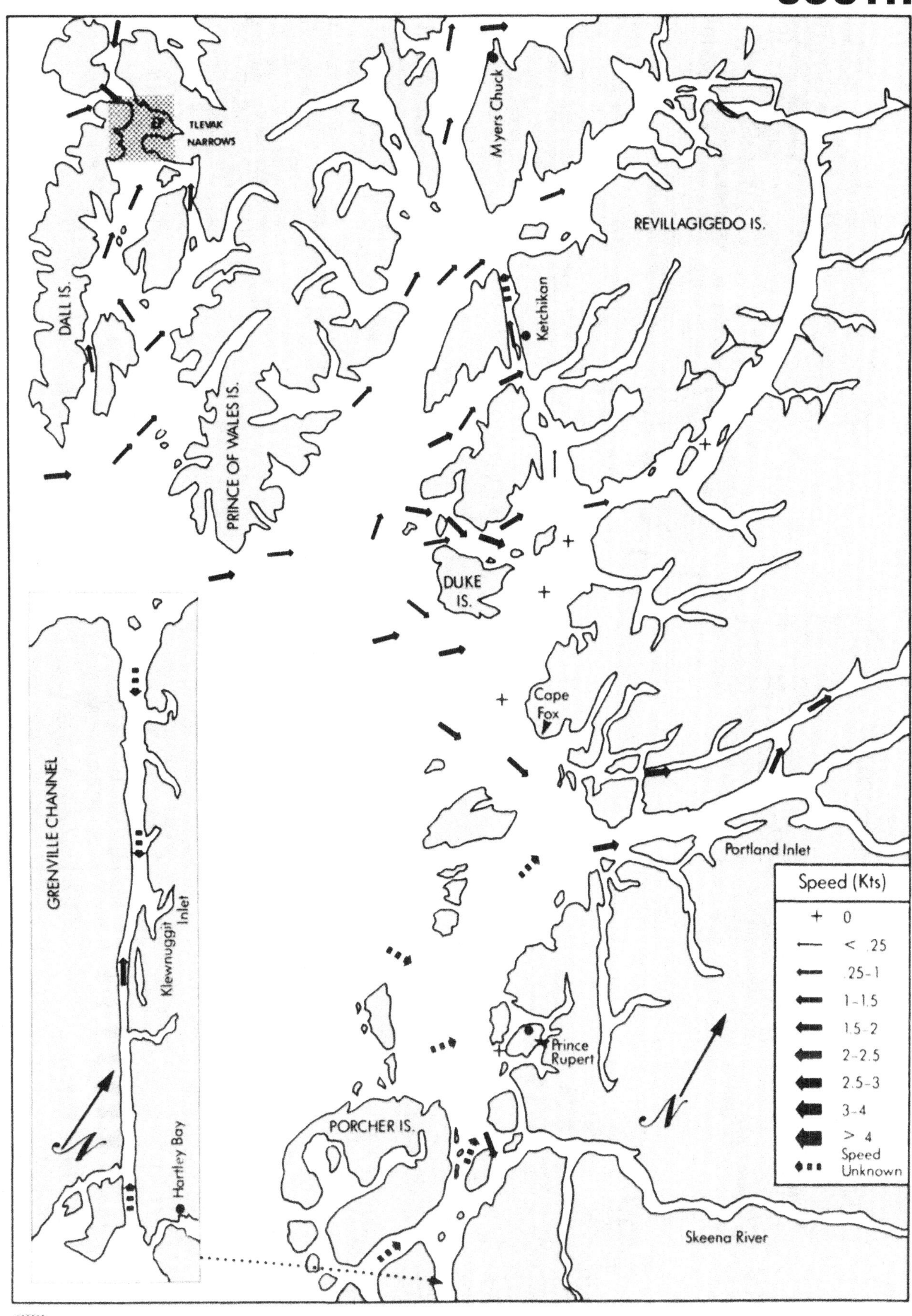
TLEVAK
NARROWS
Myers Chuck
REVILLAGIGEDO IS.
DALL IS.
Ketchikan
PRINCE OF WALES IS.
DUKE
IS.
Cape
Fox
GRENVILLE CHANNEL
Klewnuggit
Inlet
Portland Inlet
Speed (Kts)
0
< .25
.25-1
1-1.5
1.5-2
2-2.5
2.5-3
3-4
> 4
Speed
Unknown
Prince
Rupert
PORCHER IS.
Hartley Bay
Skeena River

Shaded Areas: See SELECTED CHANNELS

SOUTH

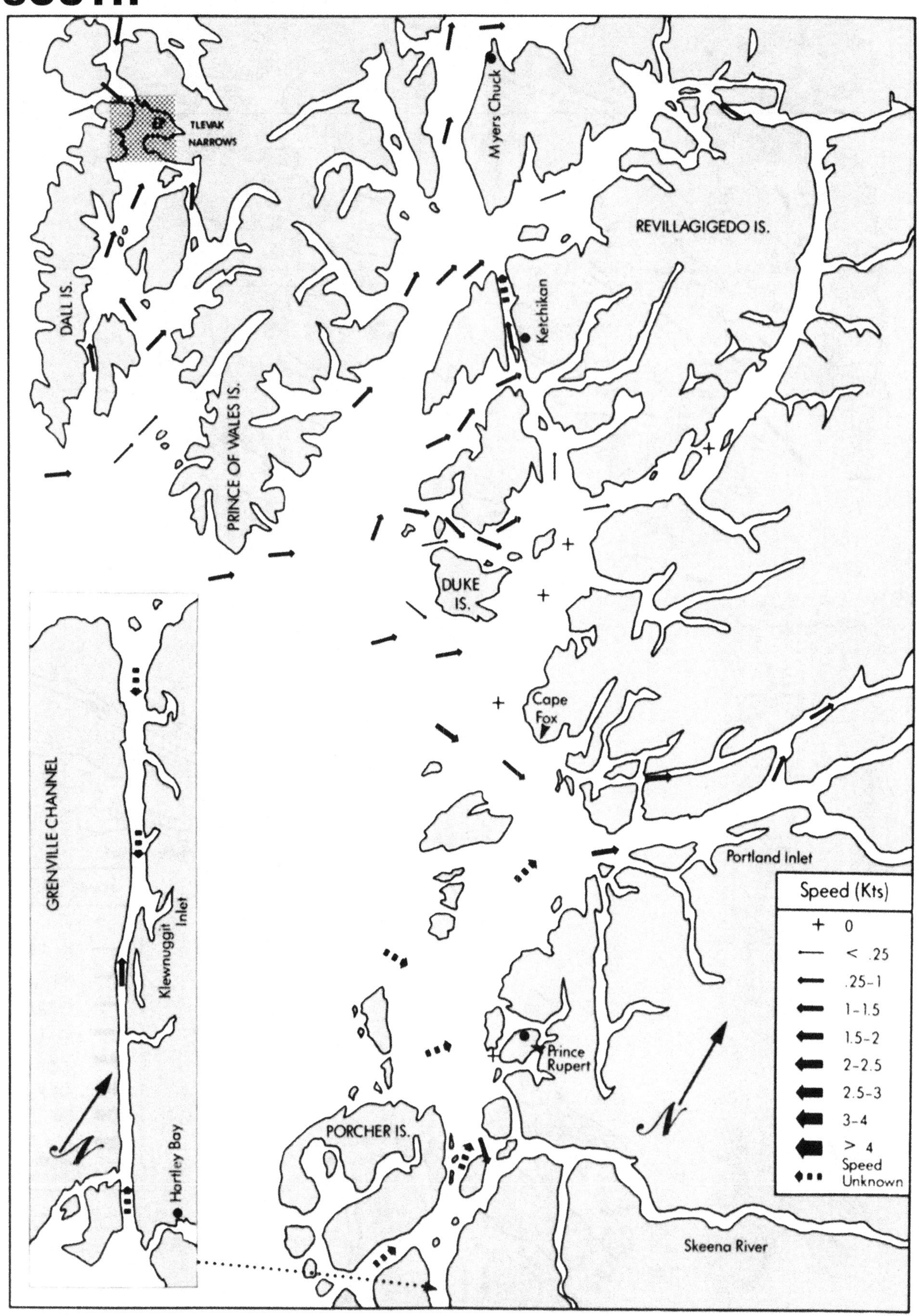

TLEVAK
NARROWS
Myers Chuck
REVILLAGIGEDO IS.
DALL IS.
Ketchikan
PRINCE OF WALES IS.
DUKE
IS.
Cape
Fox
GRENVILLE CHANNEL
Klewnuggit
Inlet
Portland Inlet
Speed (Kts)
0
< .25
.25-1
1-1.5
1.5-2
2-2.5
2.5-3
3-4
> 4
Speed
Unknown
Prince
Rupert
PORCHER IS.
Hartley Bay
Skeena River

Shaded Areas: See SELECTED CHANNELS

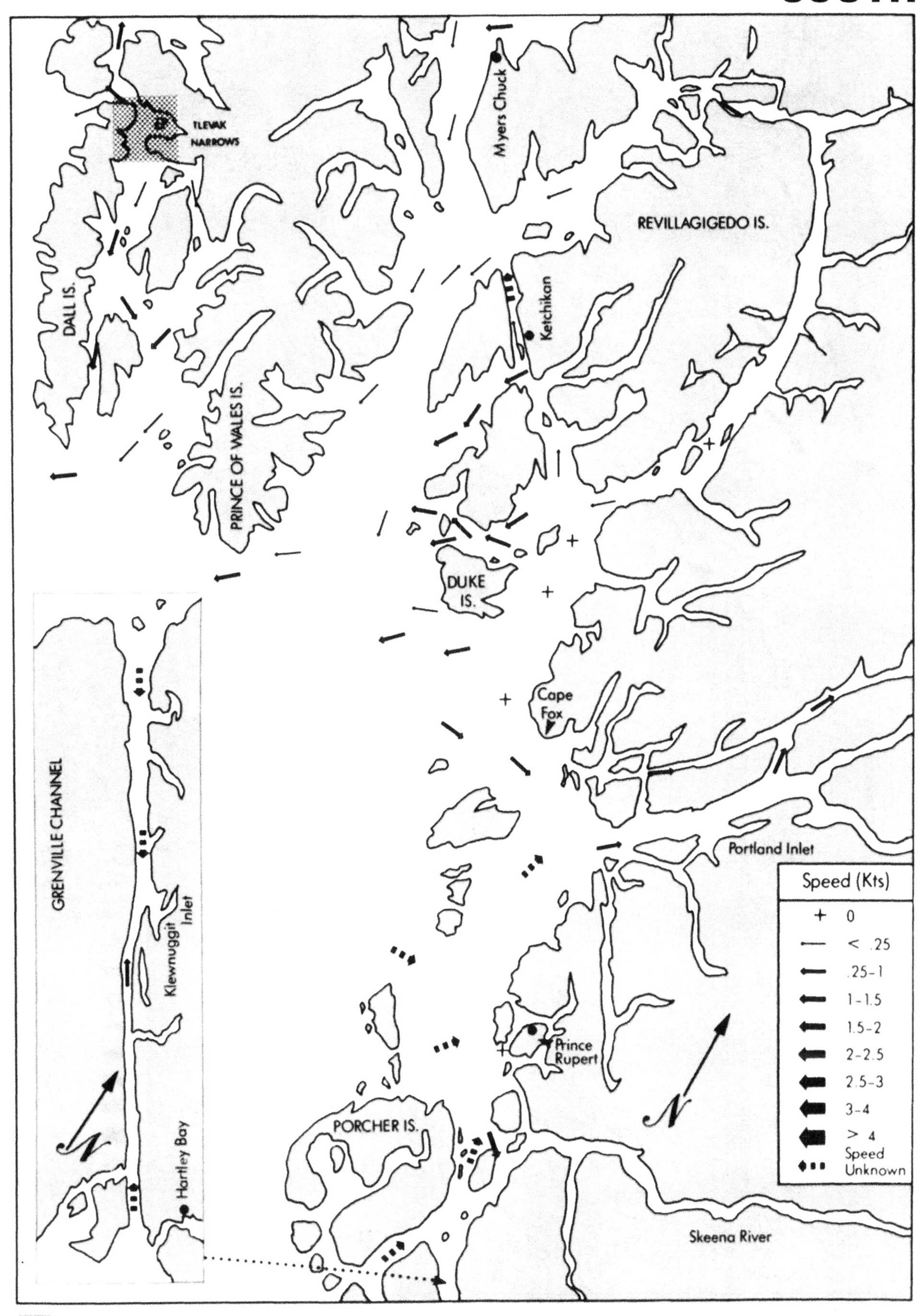
TLEVAK NARROWS
Myers Chuck
REVILLAGIGEDO IS.
DALL IS.
Ketchikan
PRINCE OF WALES IS.
DUKE IS.
Cape Fox
GRENVILLE CHANNEL
Klewnuggit Inlet
Hartley Bay
Portland Inlet
Prince Rupert
PORCHER IS.
Skeena River
Speed (Kts)
0
< .25
.25-1
1-1.5
1.5-2
2-2.5
2.5-3
3-4
> 4
Speed Unknown

Shaded Areas: See SELECTED CHANNELS

SOUTH

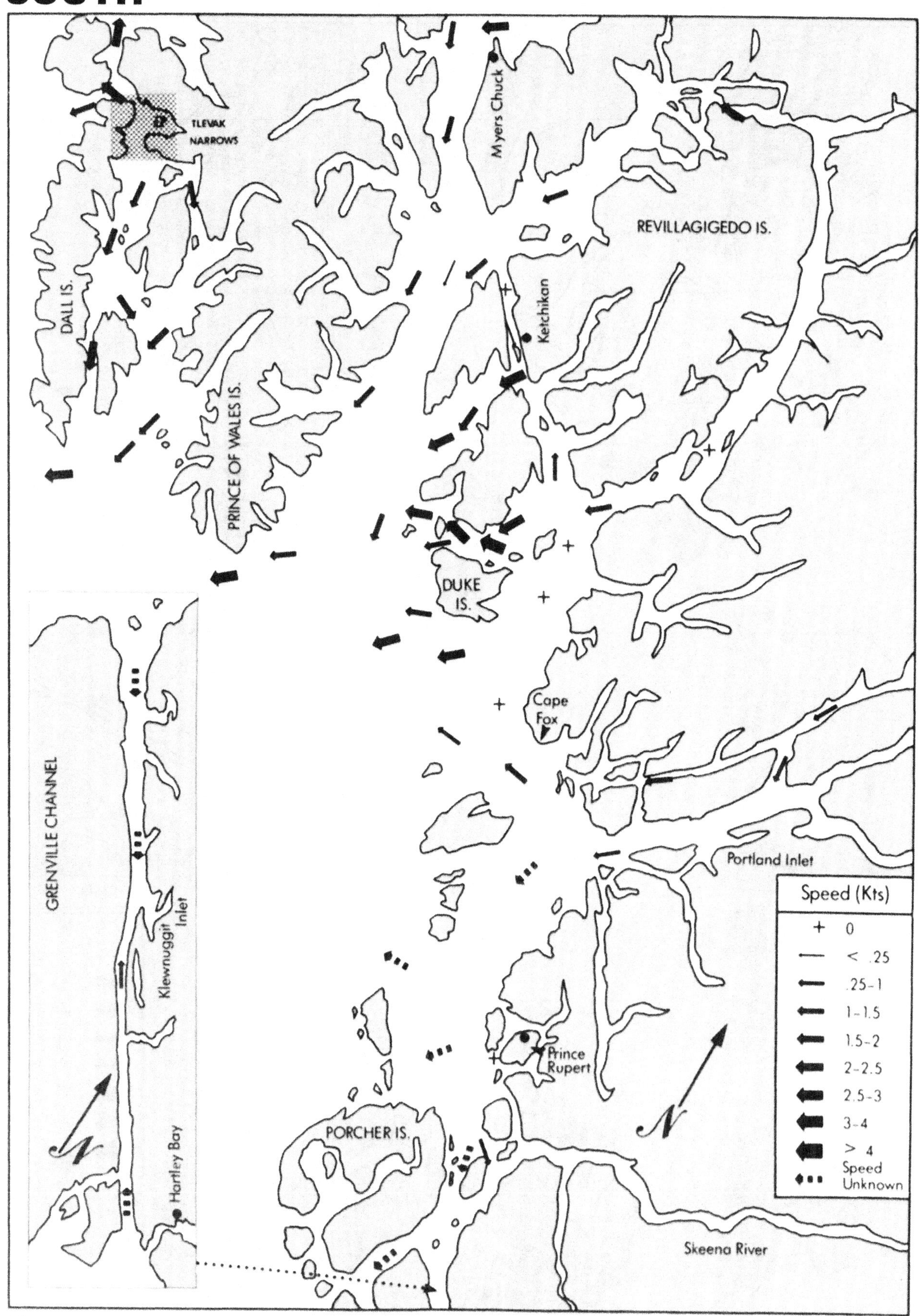

TLEVAK NARROWS
Myers Chuck
REVILLAGIGEDO IS.
DALL IS.
Ketchikan
PRINCE OF WALES IS.
DUKE IS.
Cape Fox
GRENVILLE CHANNEL
Klewnuggit Inlet
Portland Inlet
Speed (Kts)
+ 0
< .25
.25-1
1-1.5
1.5-2
2-2.5
2.5-3
3-4
> 4
Speed Unknown
Prince Rupert
PORCHER IS.
Hartley Bay
Skeena River
N
N

Shaded Areas: See SELECTED CHANNELS

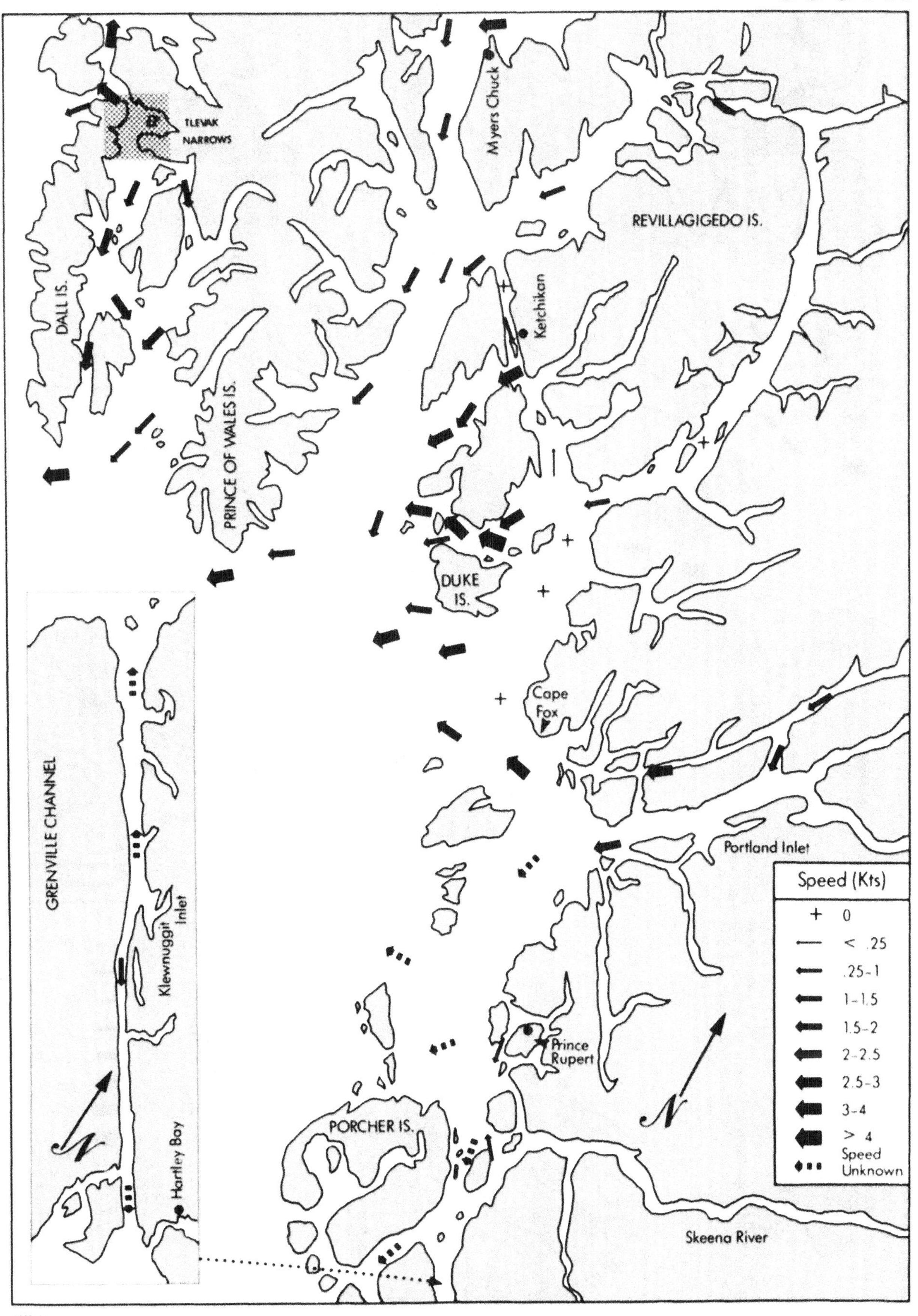
TLEVAK
NARROWS
Myers Chuck
REVILLAGIGEDO IS.
DALL IS.
Ketchikan
PRINCE OF WALES IS.
DUKE
IS.
Cape
Fox
GRENVILLE CHANNEL
Klewnuggit
Inlet
Portland Inlet
Speed (Kts)
+ 0
< .25
.25-1
1-1.5
1.5-2
2-2.5
2.5-3
3-4
> 4
Speed
Unknown
Prince
Rupert
PORCHER IS.
Hartley Bay
Skeena River

Shaded Areas: See SELECTED CHANNELS

SOUTH

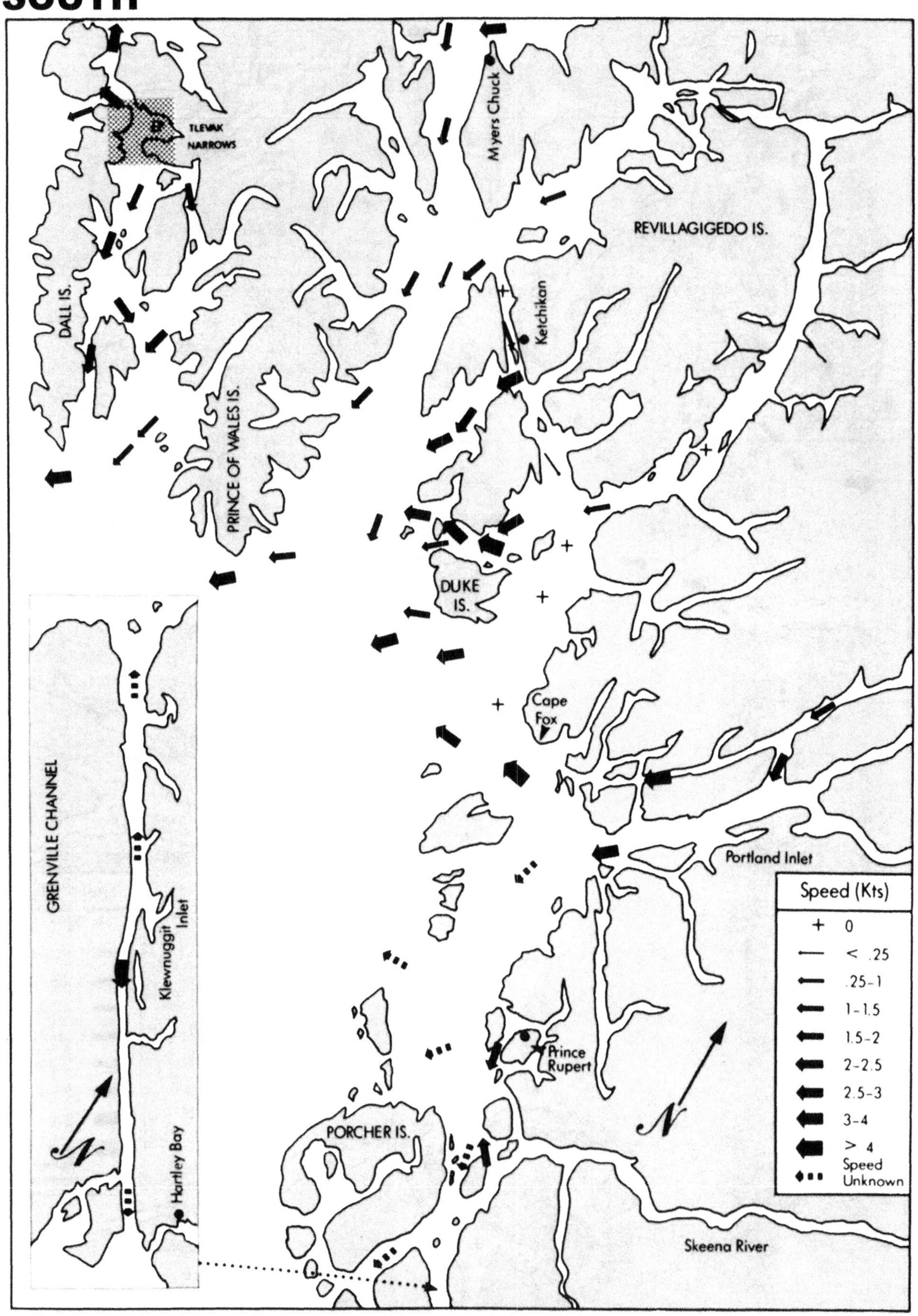

TLEVAK NARROWS
Myers Chuck
REVILLAGIGEDO IS.
DALL IS.
Ketchikan
PRINCE OF WALES IS.
DUKE IS.
Cape Fox
GRENVILLE CHANNEL
Klewnuggit Inlet
Portland Inlet
Speed (Kts)
0
< .25
.25-1
1-1.5
1.5-2
2-2.5
2.5-3
3-4
> 4
Speed Unknown
Prince Rupert
PORCHER IS.
Hartley Bay
Skeena River

Shaded Areas: See SELECTED CHANNELS

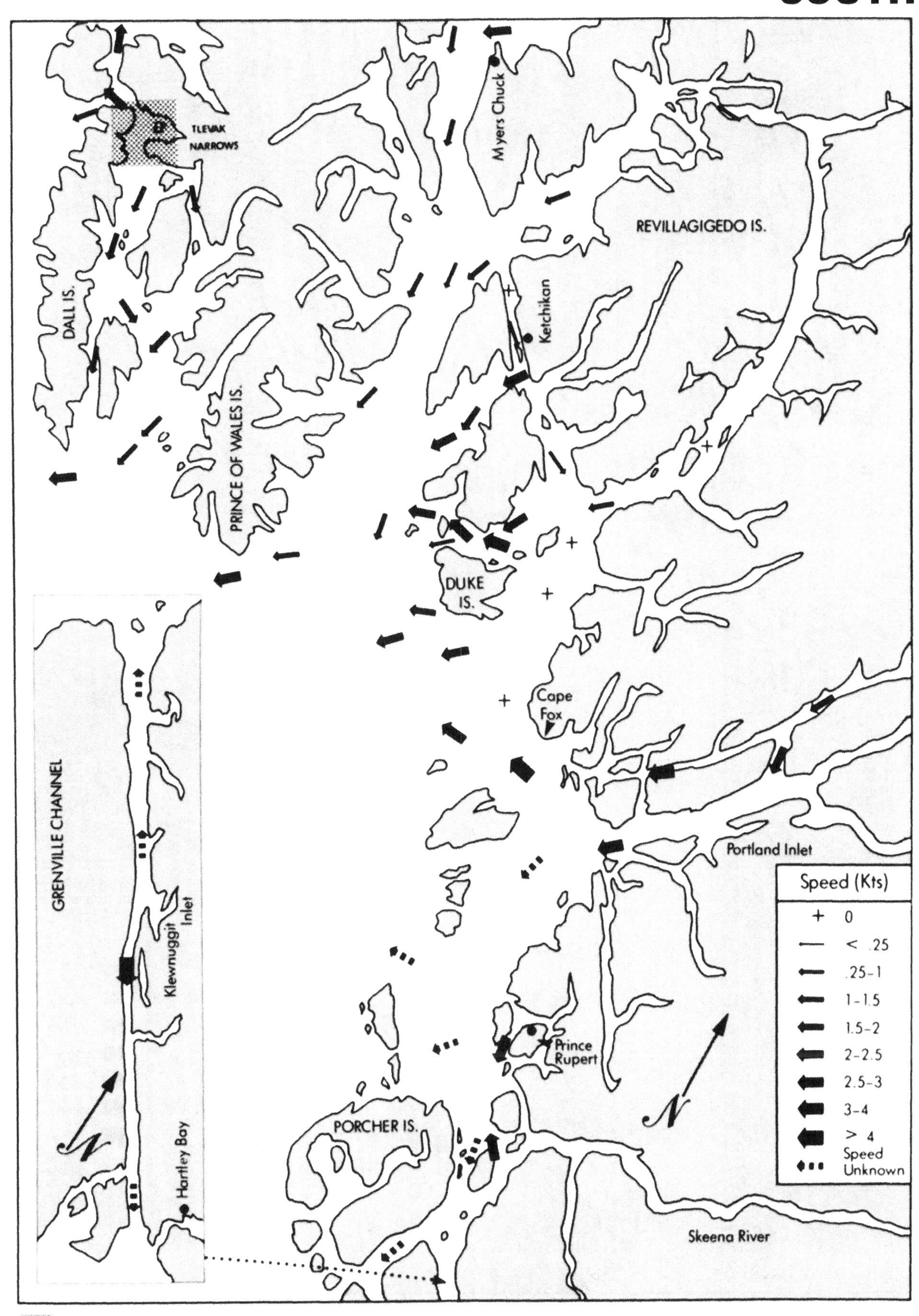
TLEVAK NARROWS
Myers Chuck
REVILLAGIGEDO IS.
Ketchikan
DALL IS.
PRINCE OF WALES IS.
DUKE IS.
Cape Fox
GRENVILLE CHANNEL
Klewnuggit Inlet
Hartley Bay
Portland Inlet
Prince Rupert
PORCHER IS.
Skeena River
Speed (Kts)
+ 0
< .25
.25-1
1-1.5
1.5-2
2-2.5
2.5-3
3-4
> 4
Speed Unknown

Shaded Areas: See SELECTED CHANNELS

SOUTH

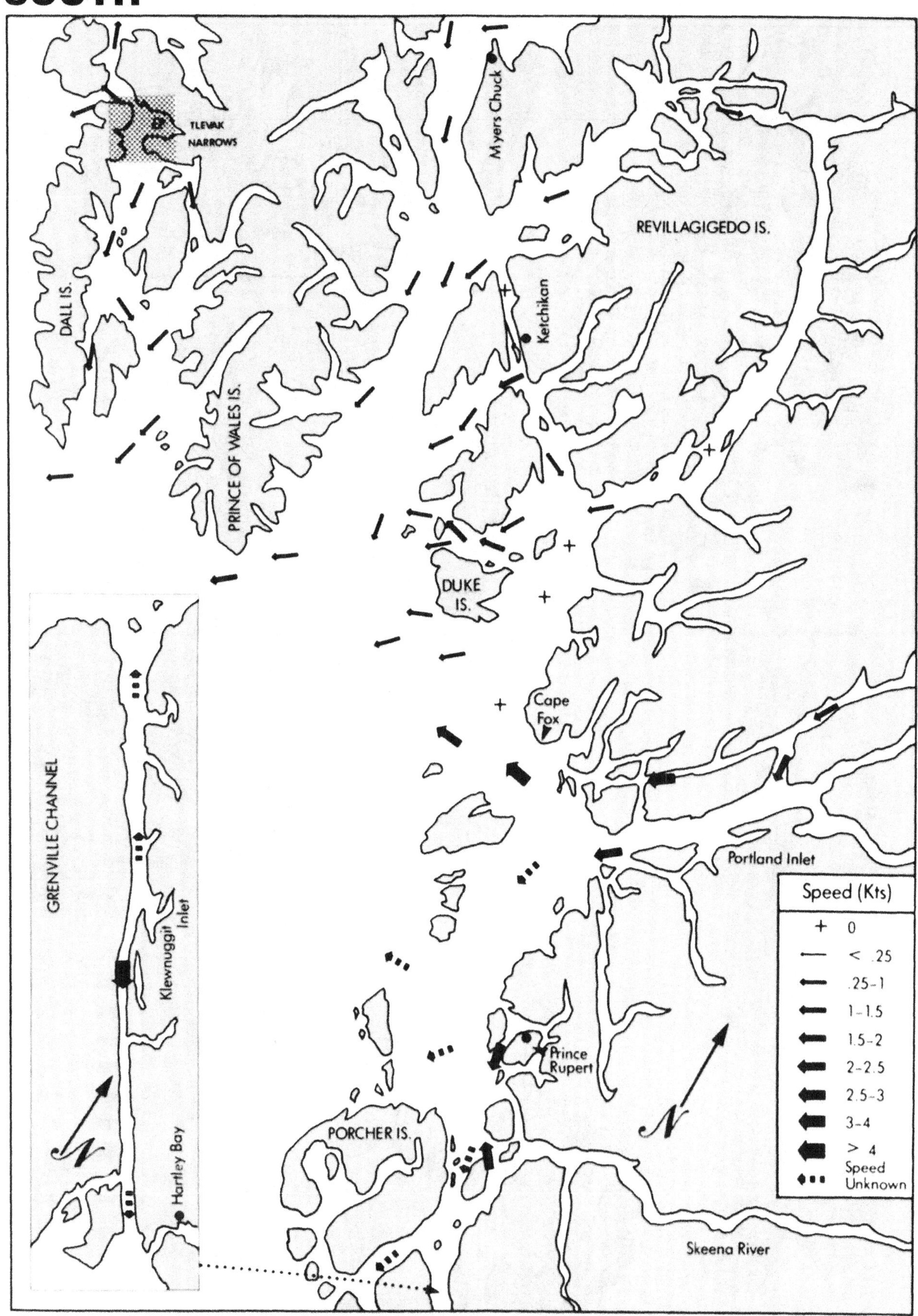

TLEVAK NARROWS
Myers Chuck
REVILLAGIGEDO IS.
DALL IS.
Ketchikan
PRINCE OF WALES IS.
DUKE IS.
Cape Fox
GRENVILLE CHANNEL
Klewnuggit Inlet
Portland Inlet
Speed (Kts)
0
< .25
.25-1
1-1.5
1.5-2
2-2.5
2.5-3
3-4
> 4
Speed Unknown
Prince Rupert
PORCHER IS.
Hartley Bay
Skeena River

Shaded Areas: See **SELECTED CHANNELS**

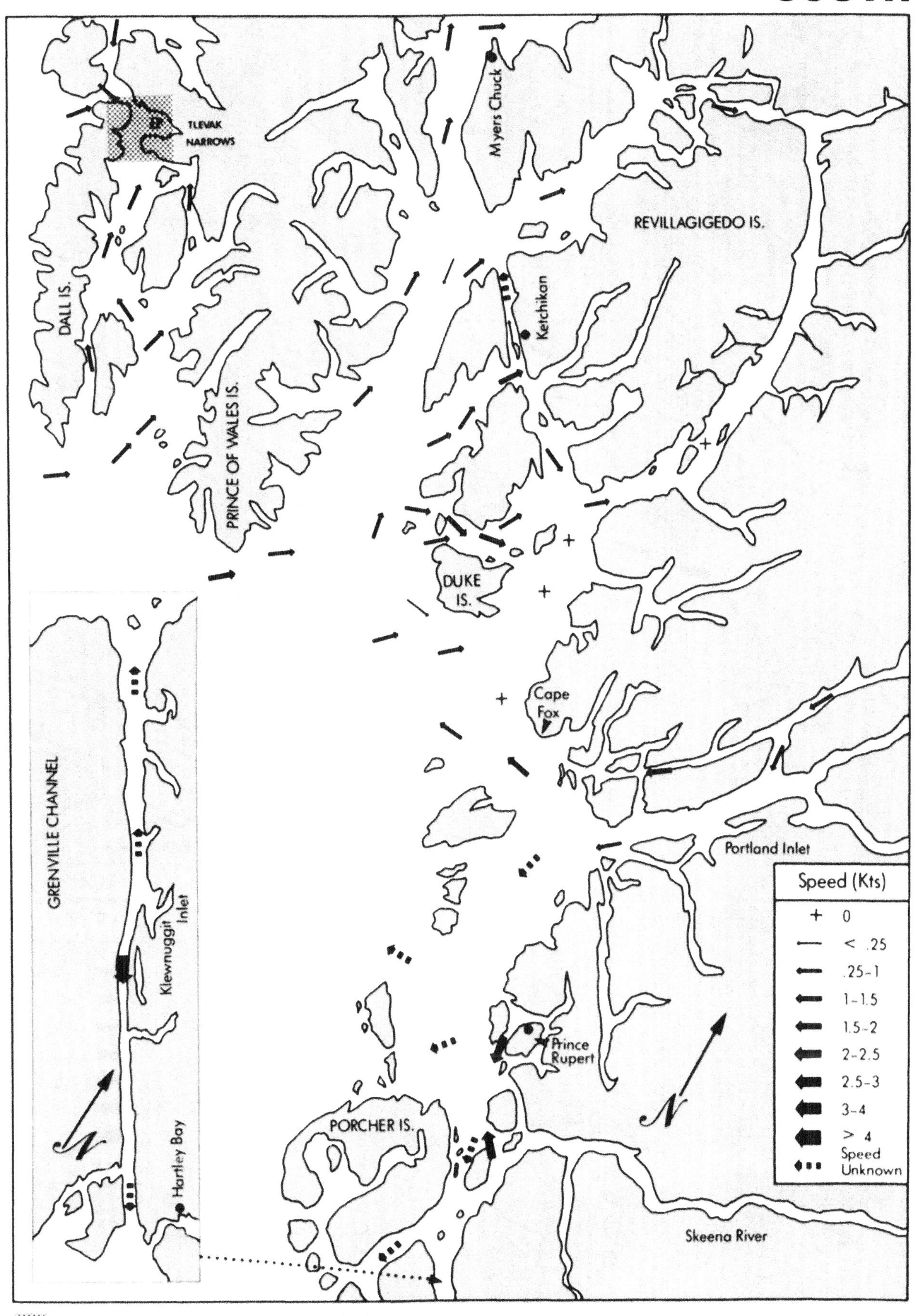
TLEVAK NARROWS
Myers Chuck
REVILLAGIGEDO IS.
DALL IS.
Ketchikan
PRINCE OF WALES IS.
DUKE IS.
Cape Fox
Portland Inlet
GRENVILLE CHANNEL
Klewnuggit Inlet
Prince Rupert
PORCHER IS.
Hartley Bay
Skeena River
Speed (Kts)
0
< .25
.25-1
1-1.5
1.5-2
2-2.5
2.5-3
3-4
> 4
Speed Unknown

Shaded Areas: See SELECTED CHANNELS

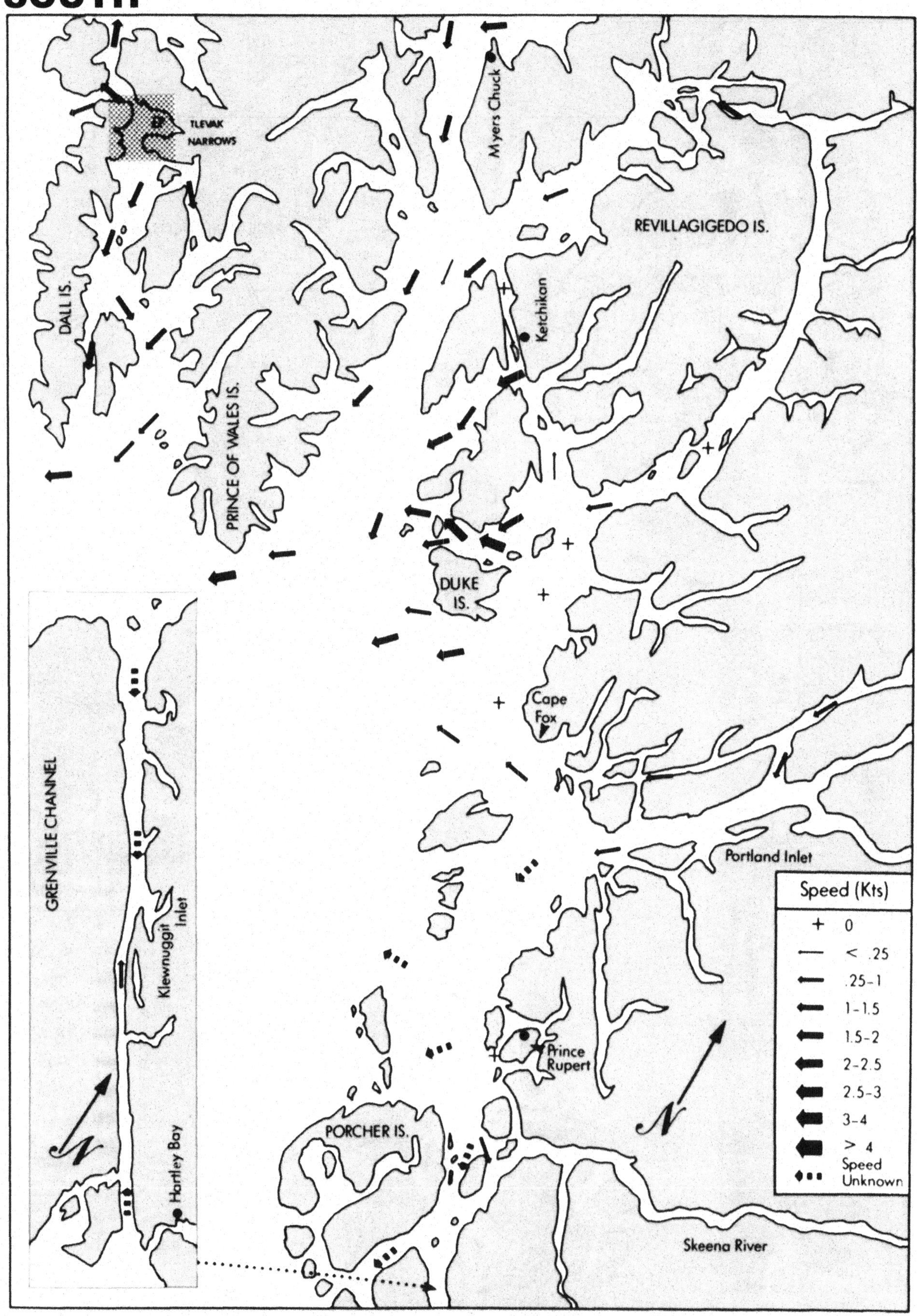
TLEVAK
NARROWS
Myers Chuck
REVILLAGIGEDO IS.
DALL IS.
Ketchikan
PRINCE OF WALES IS.
DUKE
IS.
Cape
Fox
GRENVILLE CHANNEL
Klewnuggit
Inlet
Portland Inlet
Speed (Kts)
+ 0
< .25
.25-1
1-1.5
1.5-2
2-2.5
2.5-3
3-4
> 4
Speed
Unknown
Prince
Rupert
PORCHER IS.
Hartley Bay
Skeena River

Shaded Areas: See SELECTED CHANNELS

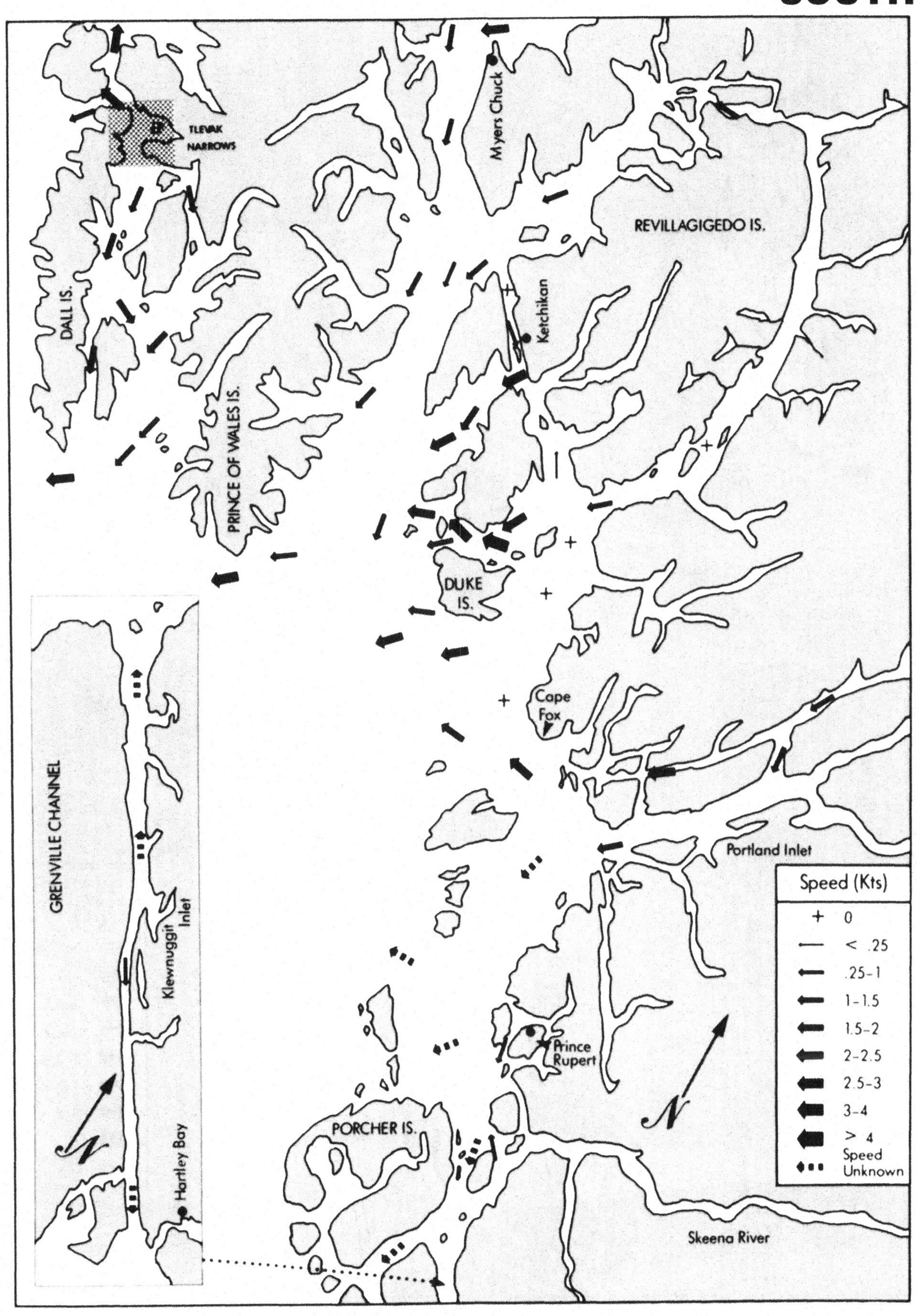
TLEVAK NARROWS
Myers Chuck
REVILLAGIGEDO IS.
DALL IS.
Ketchikan
PRINCE OF WALES IS.
DUKE IS.
Cape Fox
GRENVILLE CHANNEL
Klewnuggit Inlet
Portland Inlet
Speed (Kts)
0
< .25
.25-1
1-1.5
1.5-2
2-2.5
2.5-3
3-4
> 4
Speed Unknown
Prince Rupert
PORCHER IS.
Hartley Bay
Skeena River

Shaded Areas: See SELECTED CHANNELS

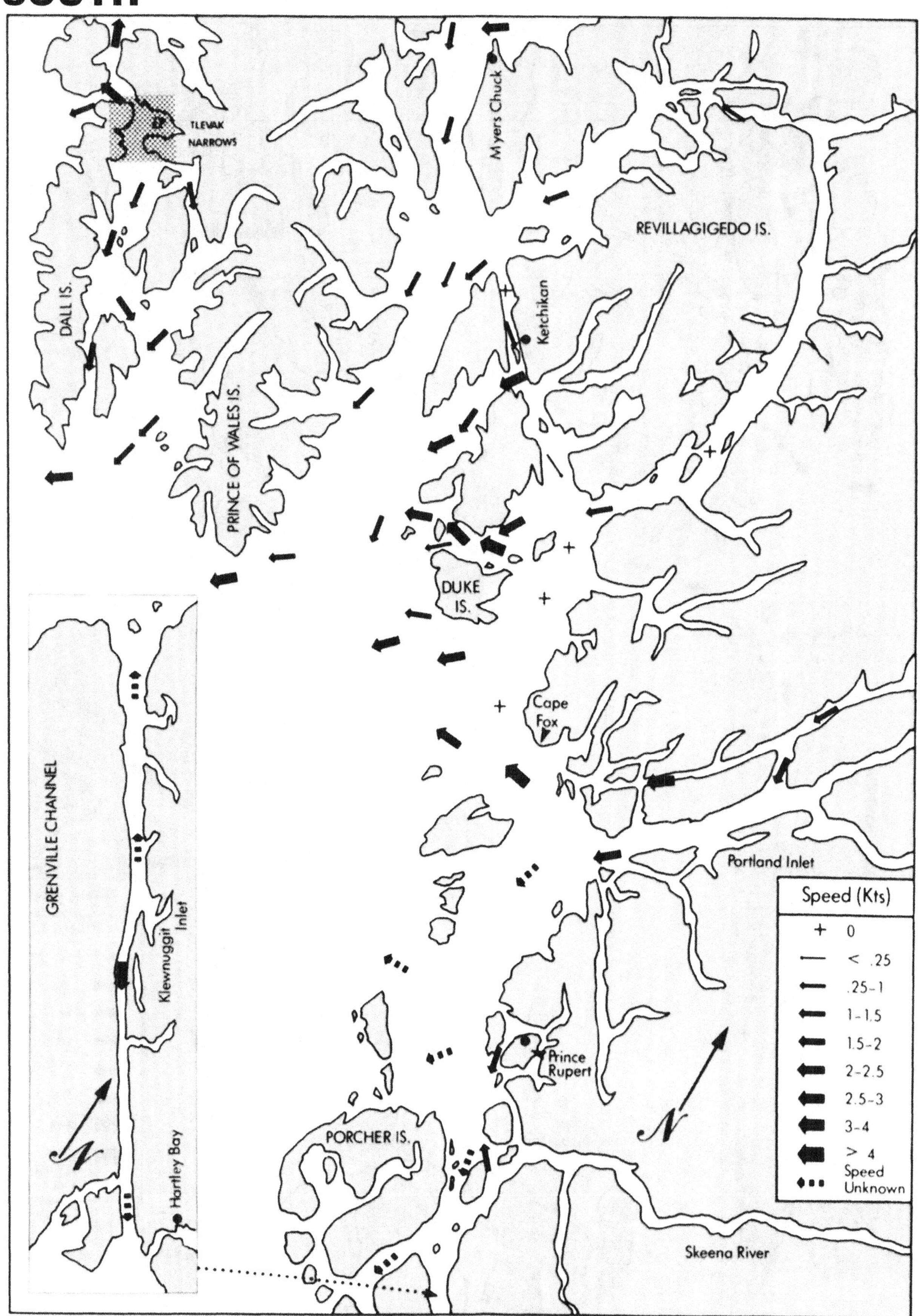
TLEVAK NARROWS
Myers Chuck
REVILLAGIGEDO IS.
DALL IS.
Ketchikan
PRINCE OF WALES IS.
DUKE IS.
Cape Fox
GRENVILLE CHANNEL
Klewnuggit Inlet
Hartley Bay
Portland Inlet
Prince Rupert
PORCHER IS.
Skeena River
Speed (Kts)
0
< .25
.25-1
1-1.5
1.5-2
2-2.5
2.5-3
3-4
> 4
Speed Unknown

Shaded Areas: See SELECTED CHANNELS

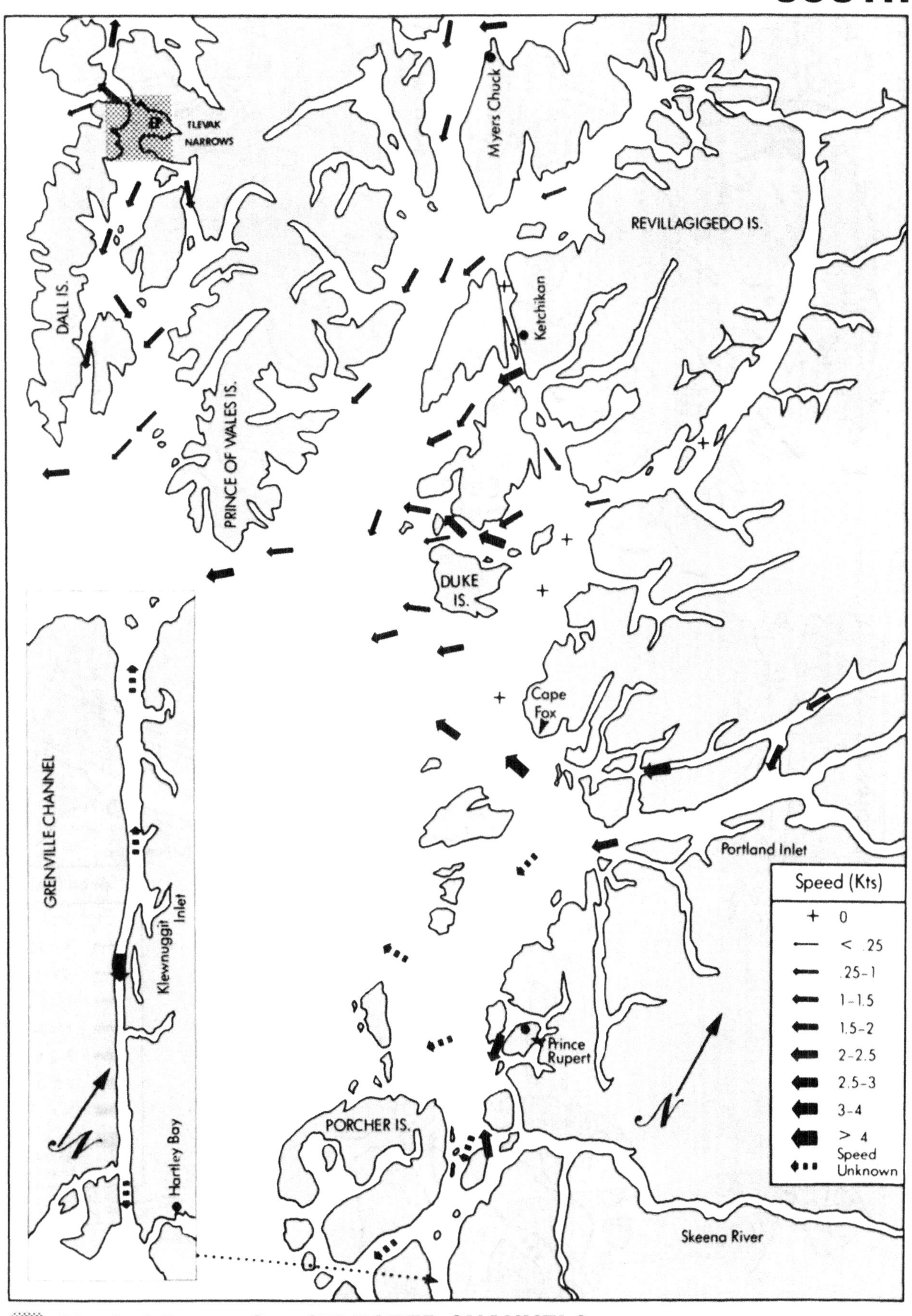
TLEVAK
NARROWS
Myers Chuck
REVILLAGIGEDO IS.
DALL IS.
Ketchikan
PRINCE OF WALES IS.
DUKE
IS.
Cape
Fox
GRENVILLE CHANNEL
Klewnuggit
Inlet
Portland Inlet
Speed (Kts)
0
< .25
.25-1
1-1.5
1.5-2
2-2.5
2.5-3
3-4
> 4
Speed
Unknown
Prince
Rupert
PORCHER IS.
Hartley Bay
Skeena River

Shaded Areas: See SELECTED CHANNELS

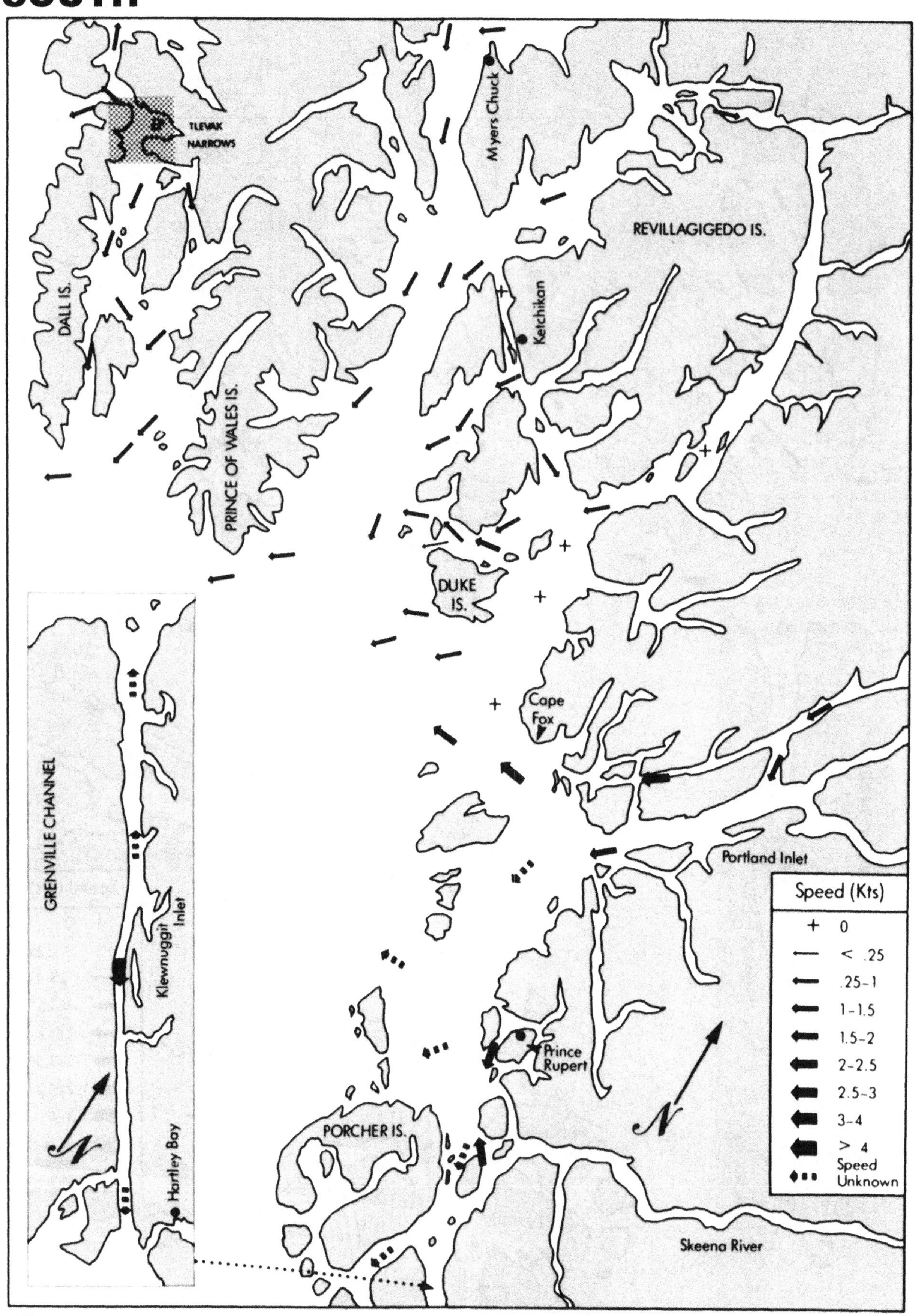
TLEVAK
NARROWS
Myers Chuck
REVILLAGIGEDO IS.
DALL IS.
Ketchikan
PRINCE OF WALES IS.
DUKE IS.
Cape Fox
GRENVILLE CHANNEL
Klewnuggit Inlet
Hartley Bay
Portland Inlet
Speed (Kts)
0
< .25
.25-1
1-1.5
1.5-2
2-2.5
2.5-3
3-4
> 4
Speed Unknown
Prince Rupert
PORCHER IS.
Skeena River

Shaded Areas: See SELECTED CHANNELS

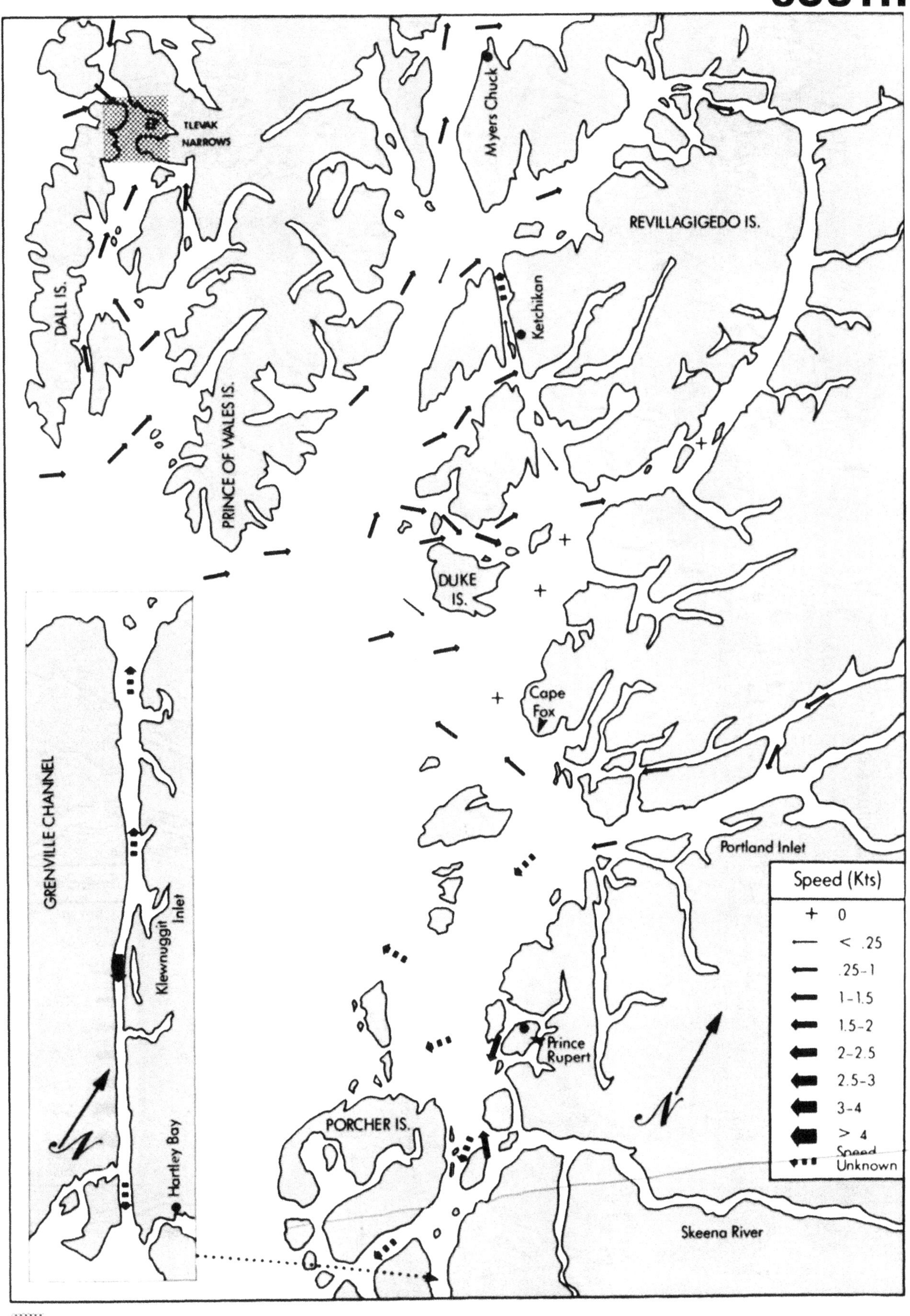
TLEVAK NARROWS
Myers Chuck
REVILLAGIGEDO IS.
DALL IS.
Ketchikan
PRINCE OF WALES IS.
DUKE IS.
Cape Fox
GRENVILLE CHANNEL
Klewnuggit Inlet
Hartley Bay
Portland Inlet
Speed (Kts)
+ 0
< .25
.25-1
1-1.5
1.5-2
2-2.5
2.5-3
3-4
> 4
Speed Unknown
Prince Rupert
PORCHER IS.
Skeena River

Shaded Areas: See SELECTED CHANNELS

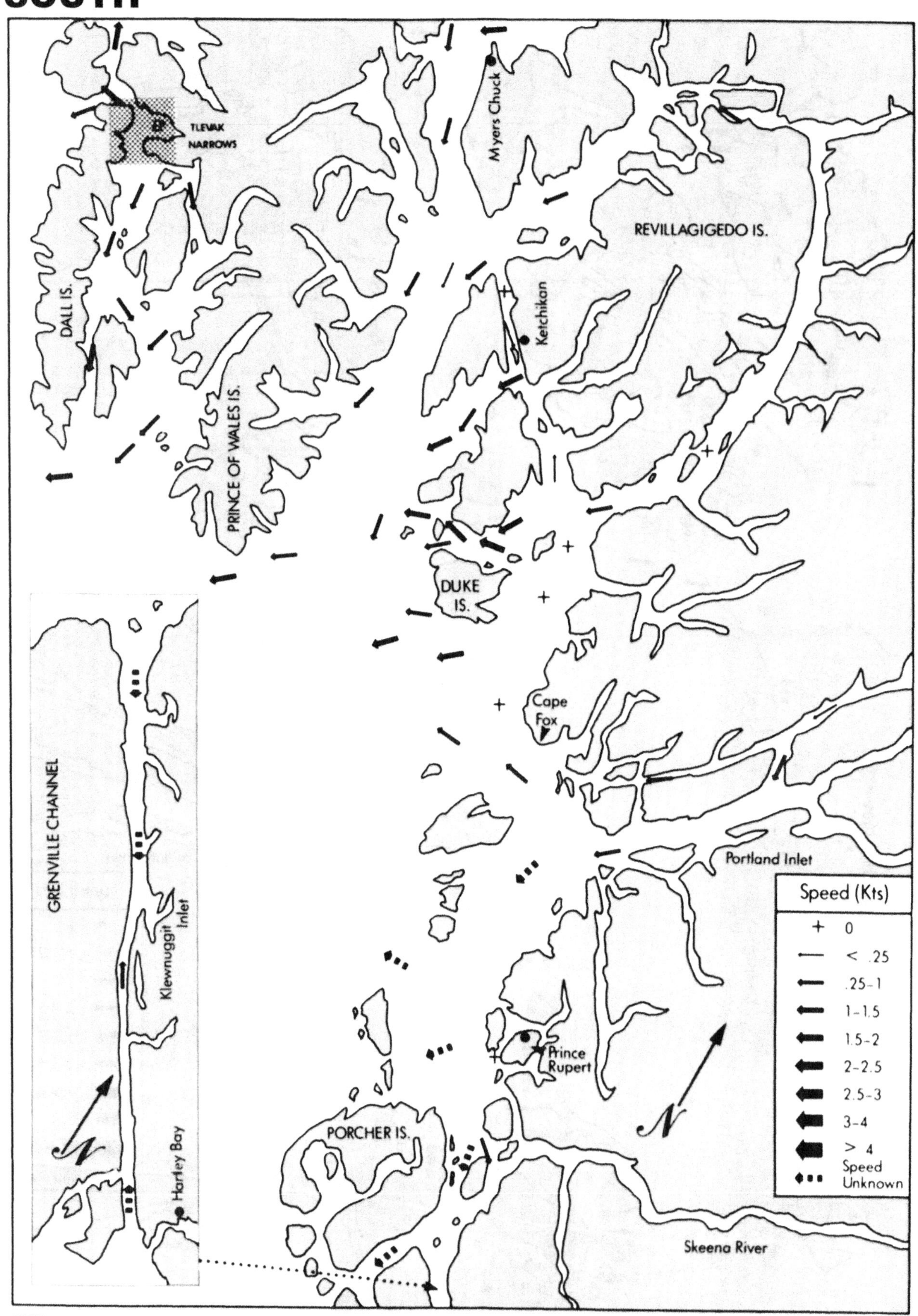
TLEVAK
NARROWS
Myers Chuck
REVILLAGIGEDO IS.
DALL IS.
Ketchikan
PRINCE OF WALES IS.
DUKE
IS.
Cape
Fox
GRENVILLE CHANNEL
Klewnuggit
Inlet
Portland Inlet
Speed (Kts)
0
< .25
.25-1
1-1.5
1.5-2
2-2.5
2.5-3
3-4
> 4
Speed
Unknown
Prince
Rupert
PORCHER IS.
Hartley Bay
Skeena River

Shaded Areas: See SELECTED CHANNELS

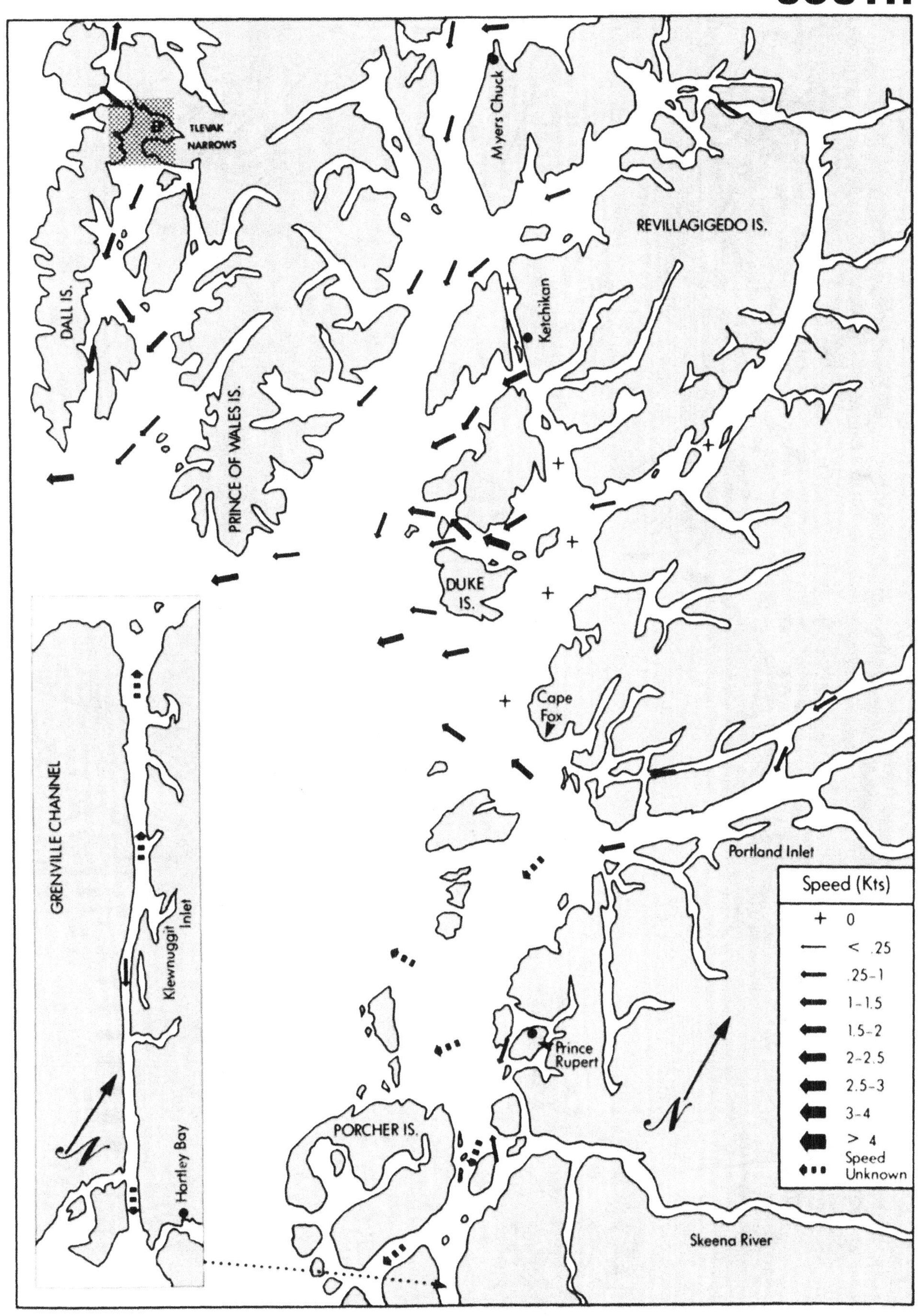
TLEVAK NARROWS
Myers Chuck
REVILLAGIGEDO IS.
DALL IS.
PRINCE OF WALES IS.
Ketchikan
DUKE IS.
Cape Fox
Portland Inlet
GRENVILLE CHANNEL
Klewnuggit Inlet
Hartley Bay
Prince Rupert
PORCHER IS.
Skeena River
Speed (Kts)
0
< .25
.25-1
1-1.5
1.5-2
2-2.5
2.5-3
3-4
> 4
Speed Unknown

Shaded Areas: See SELECTED CHANNELS

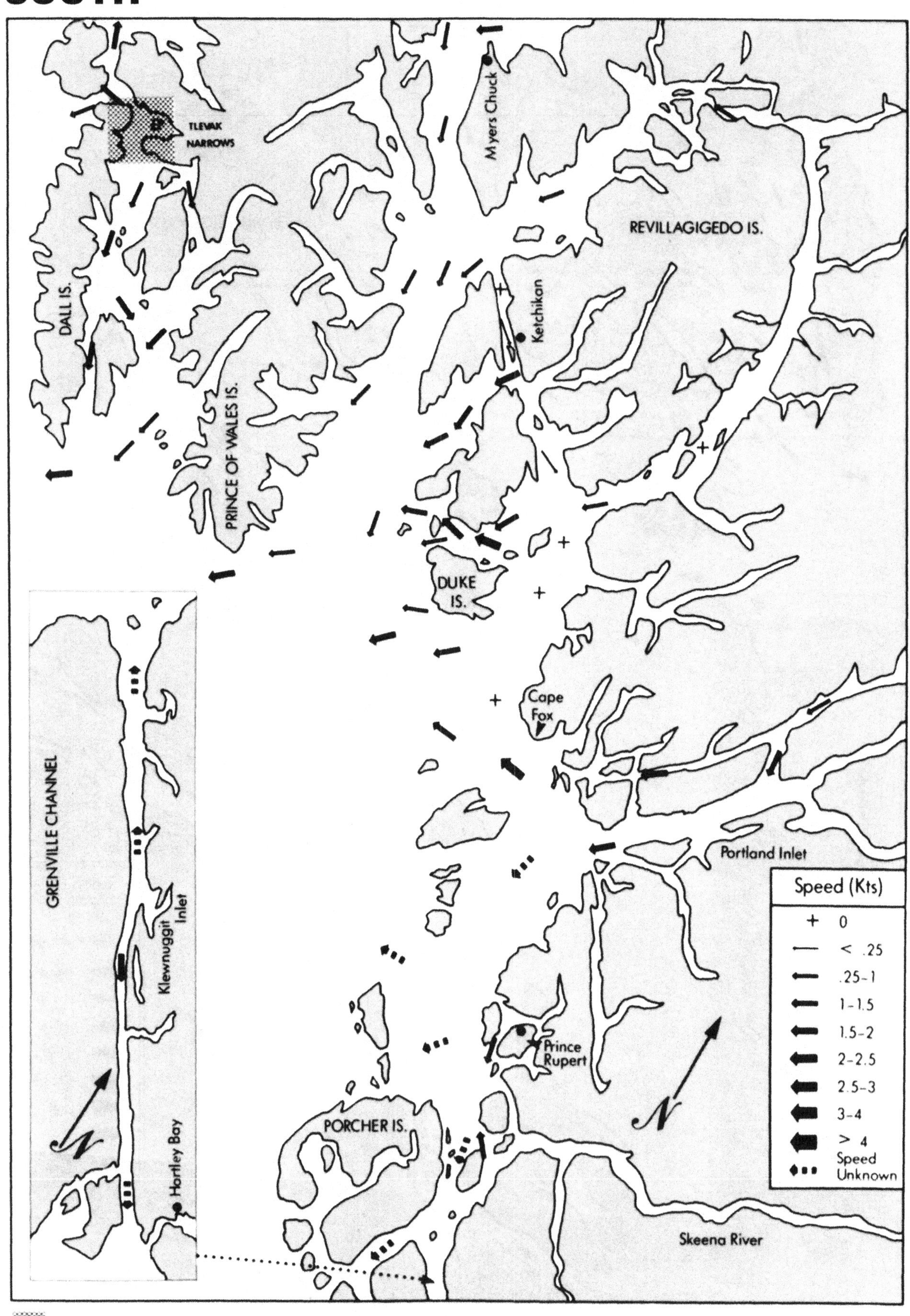
TLEVAK
NARROWS
Myers Chuck
REVILLAGIGEDO IS.
DALL IS.
Ketchikan
PRINCE OF WALES IS.
DUKE
IS.
Cape
Fox
GRENVILLE CHANNEL
Klewnuggit
Inlet
Portland Inlet
Speed (Kts)
0
< .25
.25-1
1-1.5
1.5-2
2-2.5
2.5-3
3-4
> 4
Speed
Unknown
Prince
Rupert
PORCHER IS.
Hartley Bay
Skeena River

Shaded Areas: See SELECTED CHANNELS

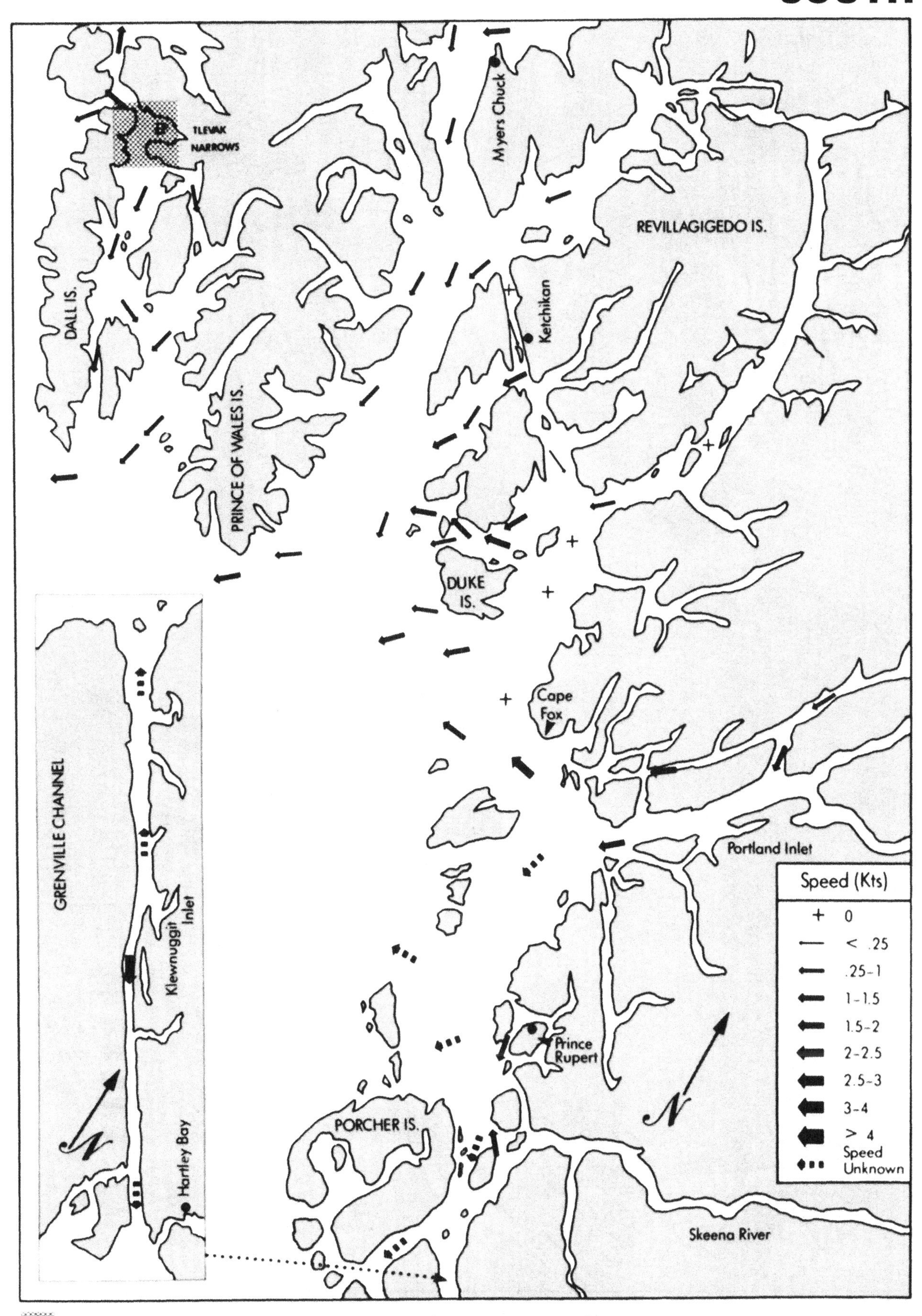
TLEVAK NARROWS
Myers Chuck
REVILLAGIGEDO IS.
DALL IS.
Ketchikan
PRINCE OF WALES IS.
DUKE IS.
Cape Fox
Portland Inlet
GRENVILLE CHANNEL
Klewnuggit Inlet
Hartley Bay
Prince Rupert
PORCHER IS.
Skeena River
Speed (Kts)
0
< .25
.25-1
1-1.5
1.5-2
2-2.5
2.5-3
3-4
> 4
Speed Unknown

Shaded Areas: See SELECTED CHANNELS

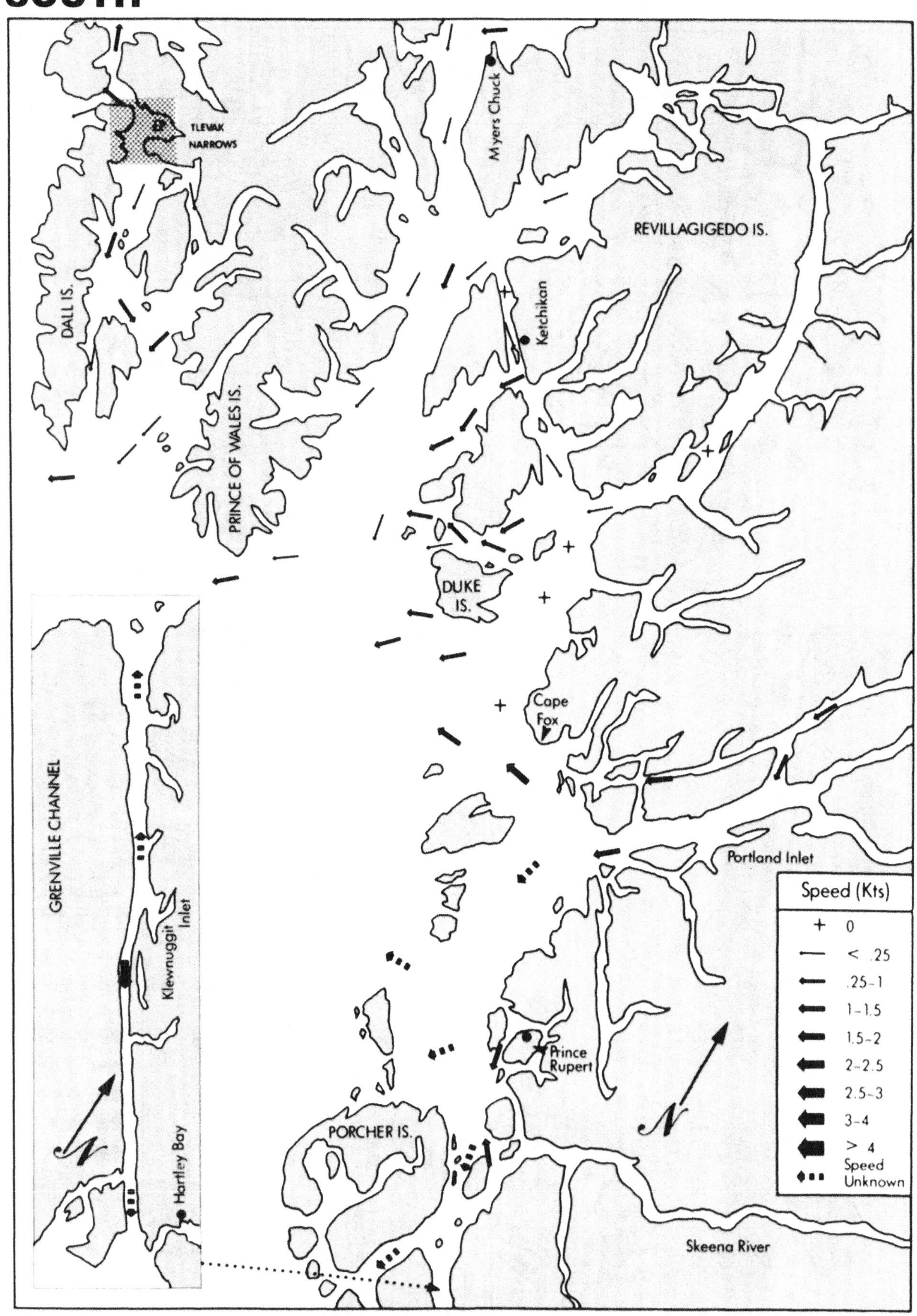
TLEVAK NARROWS
Myers Chuck
REVILLAGIGEDO IS.
DALL IS.
Ketchikan
PRINCE OF WALES IS.
DUKE IS.
Cape Fox
GRENVILLE CHANNEL
Klewnuggit Inlet
Hartley Bay
Portland Inlet
Speed (Kts)
0
< .25
.25-1
1-1.5
1.5-2
2-2.5
2.5-3
3-4
> 4
Speed Unknown
Prince Rupert
PORCHER IS.
Skeena River

Shaded Areas: See SELECTED CHANNELS

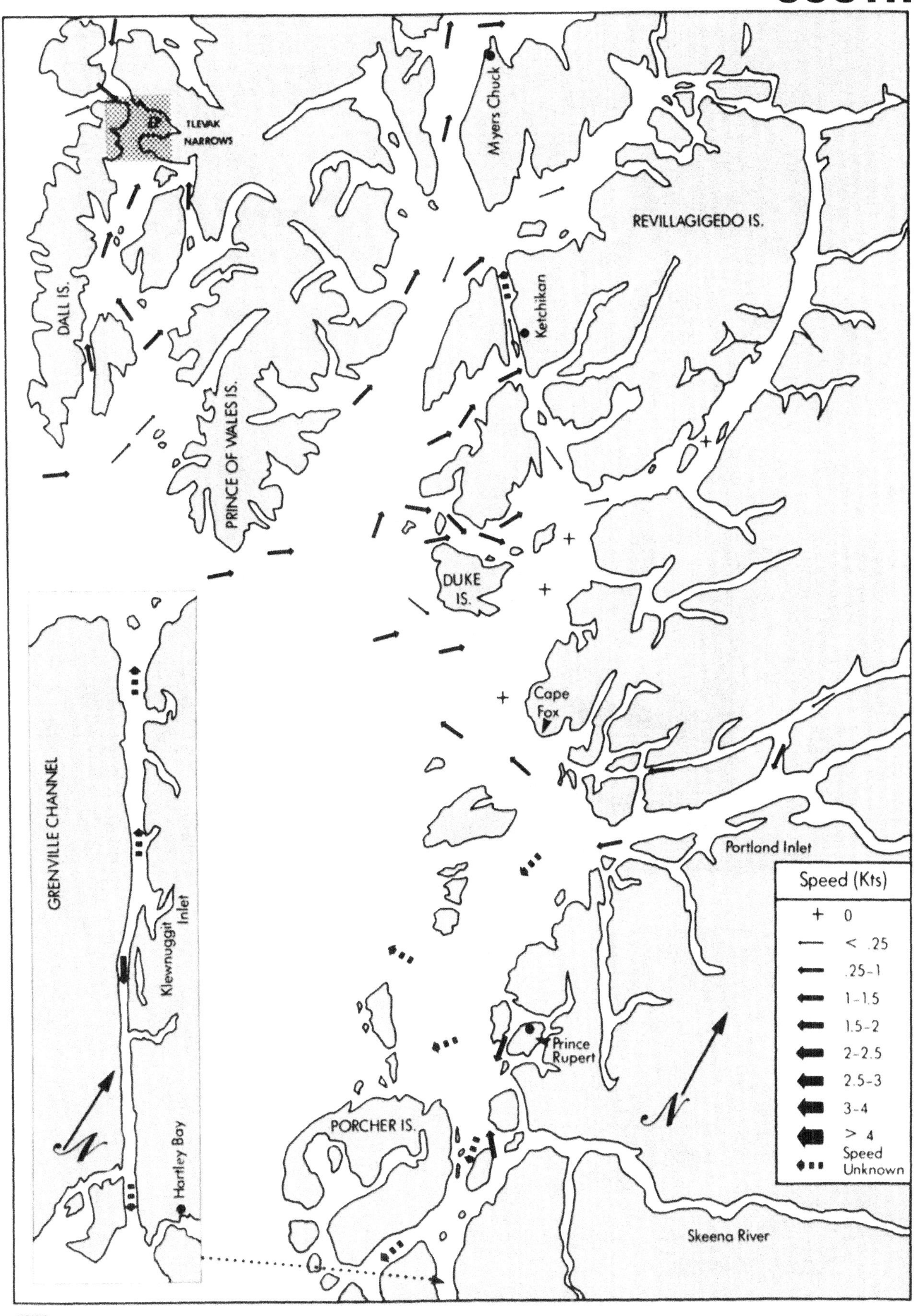
TLEVAK NARROWS
Myers Chuck
REVILLAGIGEDO IS.
DALL IS.
Ketchikan
PRINCE OF WALES IS.
DUKE IS.
Cape Fox
GRENVILLE CHANNEL
Portland Inlet
Speed (Kts)
0
< .25
.25-1
1-1.5
1.5-2
2-2.5
2.5-3
3-4
> 4
Speed Unknown
Klewnuggit Inlet
Prince Rupert
PORCHER IS.
Hartley Bay
Skeena River

Shaded Areas: See SELECTED CHANNELS

Washburne

SELECTED CHANNELS INSTRUCTIONS

Follow this procedure to determine which chart to use for a specific time and date.

Step 1. Find the Sergius Narrows current table entries for the day in question. In the left-hand column, find the time for slack water preceding or coinciding with the hour of interest. Round this slack time to the nearest hour.

Step 2. Find the maximum speed of the current following the slack determined in Step 1. This speed will be on the same line, at the right edge of the day's table—or on the first line of the next day's table if the slack occurs at the end of the day. Note whether the current is a Flood (F) or an Ebb (E).

Step 3. Using the maximum current found in Step 2, determine which row of Table 1 to use for finding the chart number. For example, a Sergius Narrows maximum Ebb of 5.5 knots would use the row marked "5.2 – 5.7" in the Ebb column.

Step 4. Compute the number of hours that elapsed since the slack found in Step 1. This will determine the column to use to find the correct chart.

For example, if the time of interest is 1500 and the most recent slack was at 1300, the elapsed time is two hours. Read across the row found in Step 3 to find the chart recommended for the number of hours since most recent slack.

In other words, if the time of interest is 1500, and this time is two hours later than the most recent slack at Sergius Narrows, which is building to a maximum of 5.5 kts Ebb, then Table 2 tells us that 1500 corresponds to chart no. 27. Then from Table 2 we know immediately that 1600 would be chart 28, and 1700 would be chart 29.

Table 2. Chart Number Based on Slack Time and Max Speed

Sergius Narrows Max. Speed (kts)	*Hours after most recent slack*					
Flood	*0h*	*1h*	*2h*	*3h*	*4h*	*5h*
> 6.3	1	2	3	4	5	6
5.7 – 6.3	7	8	9	10	11	12
< 5.7	13	14	15	16	17	18
Ebb	*0h*	*1h*	*2h*	*3h*	*4h*	*5h*
> 5.7	19	20	21	22	23	24
5.2 – 5.7	25	26	27	28	29	30
< 5.2	31	32	33	34	35	36

SELECTED CHANNELS

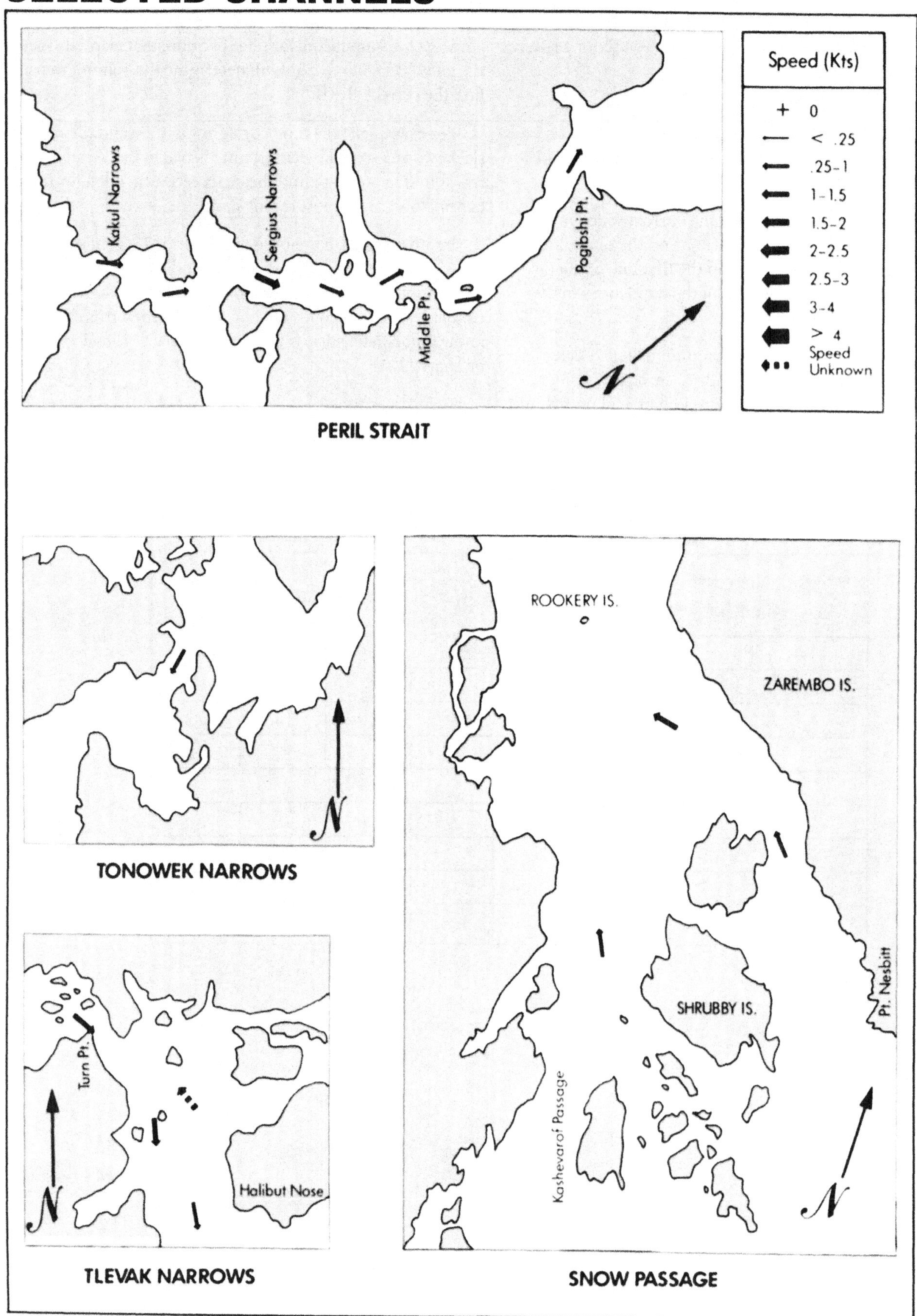
Speed (Kts)
+ 0
< .25
.25-1
1-1.5
1.5-2
2-2.5
2.5-3
3-4
> 4
Speed Unknown
Kakul Narrows
Sergius Narrows
Middle Pt.
Pogibshi Pt.
N
PERIL STRAIT
TONOWEK NARROWS
N
Turn Pt.
Halibut Nose
N
TLEVAK NARROWS
ROOKERY IS.
ZAREMBO IS.
SHRUBBY IS.
Pt. Nesbitt
Kashevarof Passage
N
SNOW PASSAGE

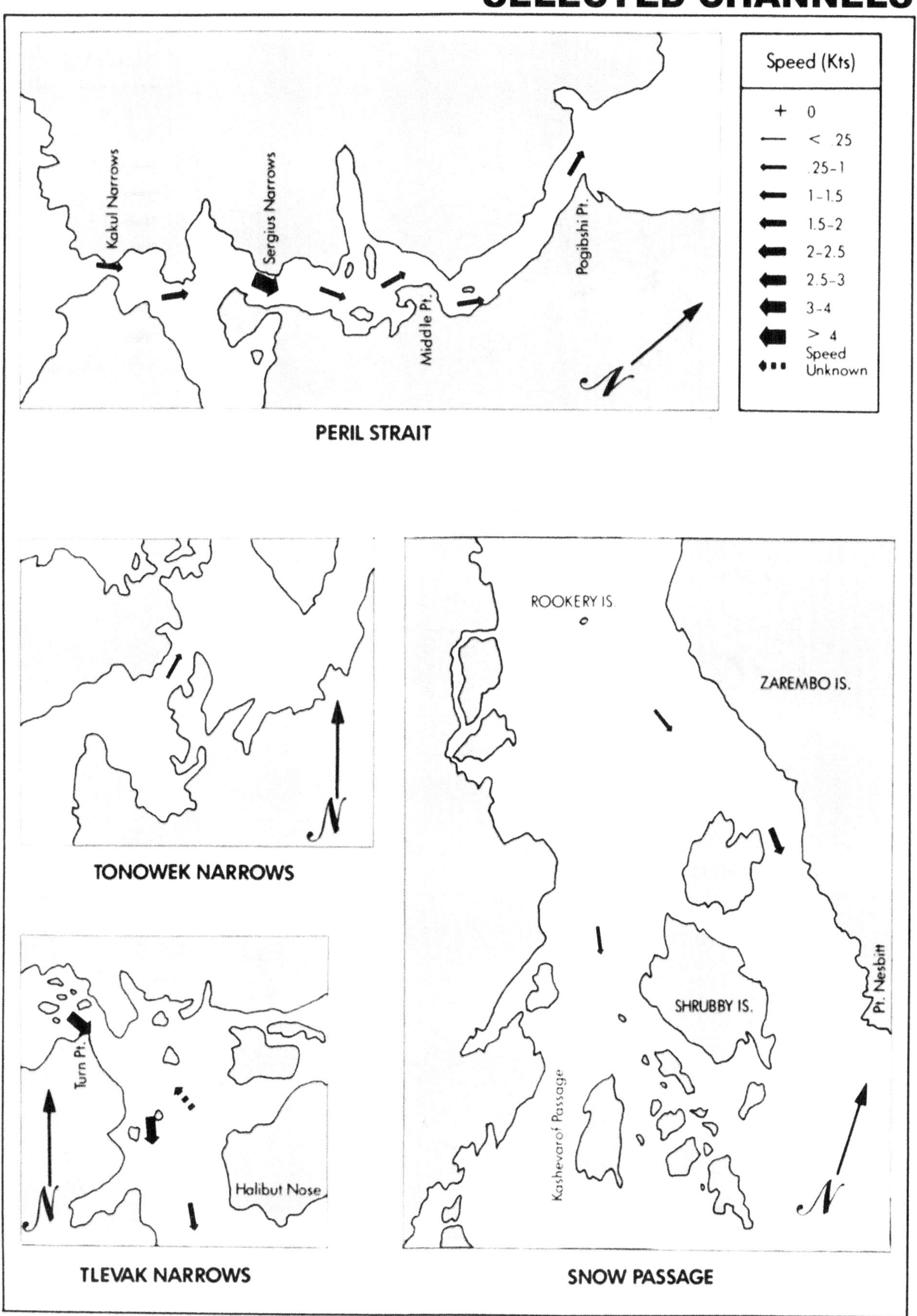
Speed (Kts)
+ 0
< .25
.25-1
1-1.5
1.5-2
2-2.5
2.5-3
3-4
> 4
Speed Unknown
Kakul Narrows
Sergius Narrows
Middle Pt.
Pogibshi Pt.
N
PERIL STRAIT
TONOWEK NARROWS
ROOKERY IS.
ZAREMBO IS.
SHRUBBY IS.
Pt. Nesbitt
Kashevarof Passage
Turn Pt.
Halibut Nose
TLEVAK NARROWS
SNOW PASSAGE

SELECTED CHANNELS

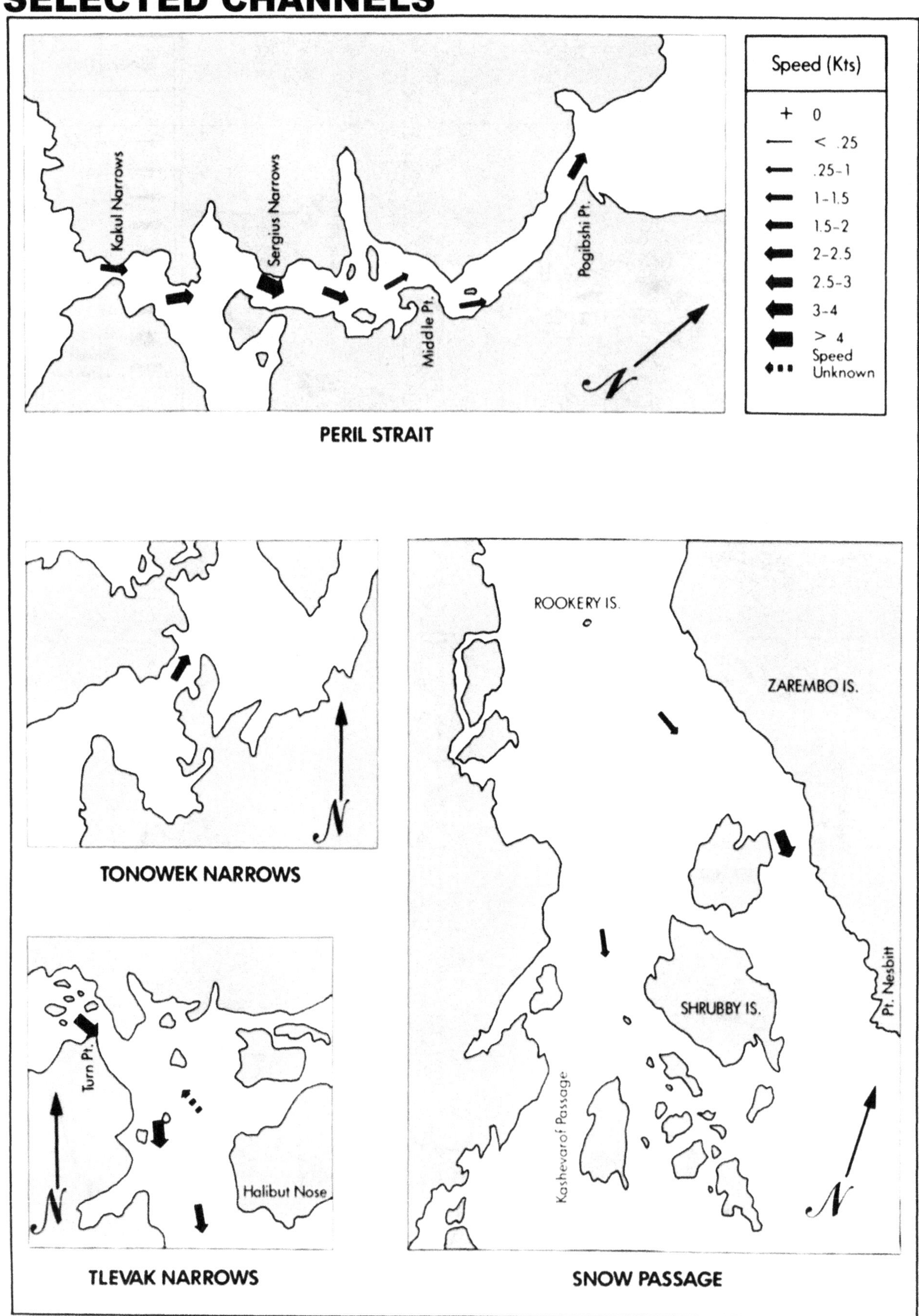
Speed (Kts)
0
< .25
.25-1
1-1.5
1.5-2
2-2.5
2.5-3
3-4
> 4
Speed Unknown
Kakul Narrows
Sergius Narrows
Middle Pt.
Pogibshi Pt.
N
PERIL STRAIT
N
TONOWEK NARROWS
Turn Pt.
Halibut Nose
N
TLEVAK NARROWS
ROOKERY IS.
ZAREMBO IS.
SHRUBBY IS.
Pt. Nesbitt
Kashevarof Passage
N
SNOW PASSAGE

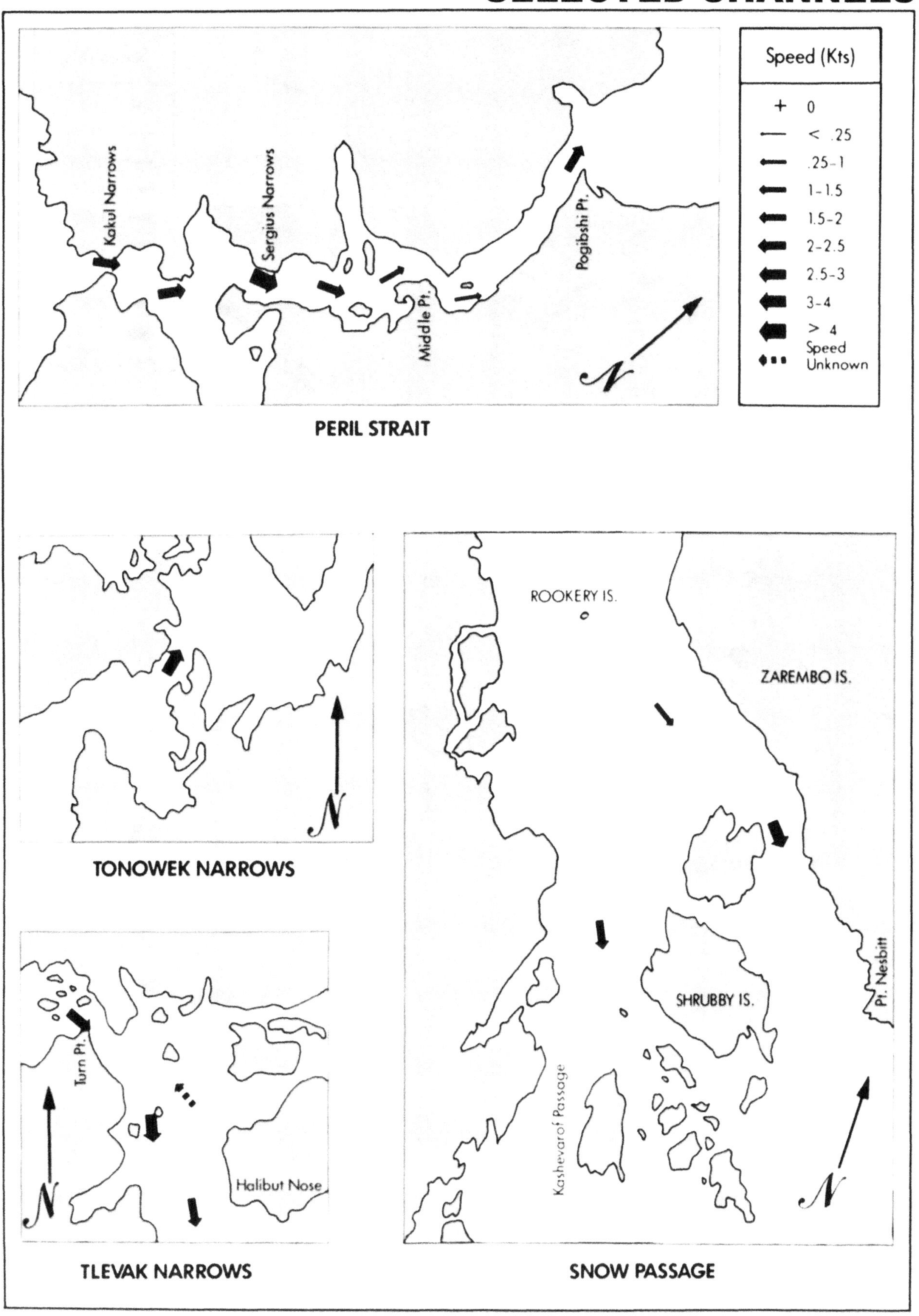
Speed (Kts)
+ 0
< .25
.25-1
1-1.5
1.5-2
2-2.5
2.5-3
3-4
> 4
Speed Unknown
Kakul Narrows
Sergius Narrows
Middle Pt.
Pogibshi Pt.
N
PERIL STRAIT
TONOWEK NARROWS
ROOKERY IS.
ZAREMBO IS.
SHRUBBY IS.
Pt. Nesbitt
Kashevarof Passage
SNOW PASSAGE
Turn Pt.
Halibut Nose
TLEVAK NARROWS

SELECTED CHANNELS

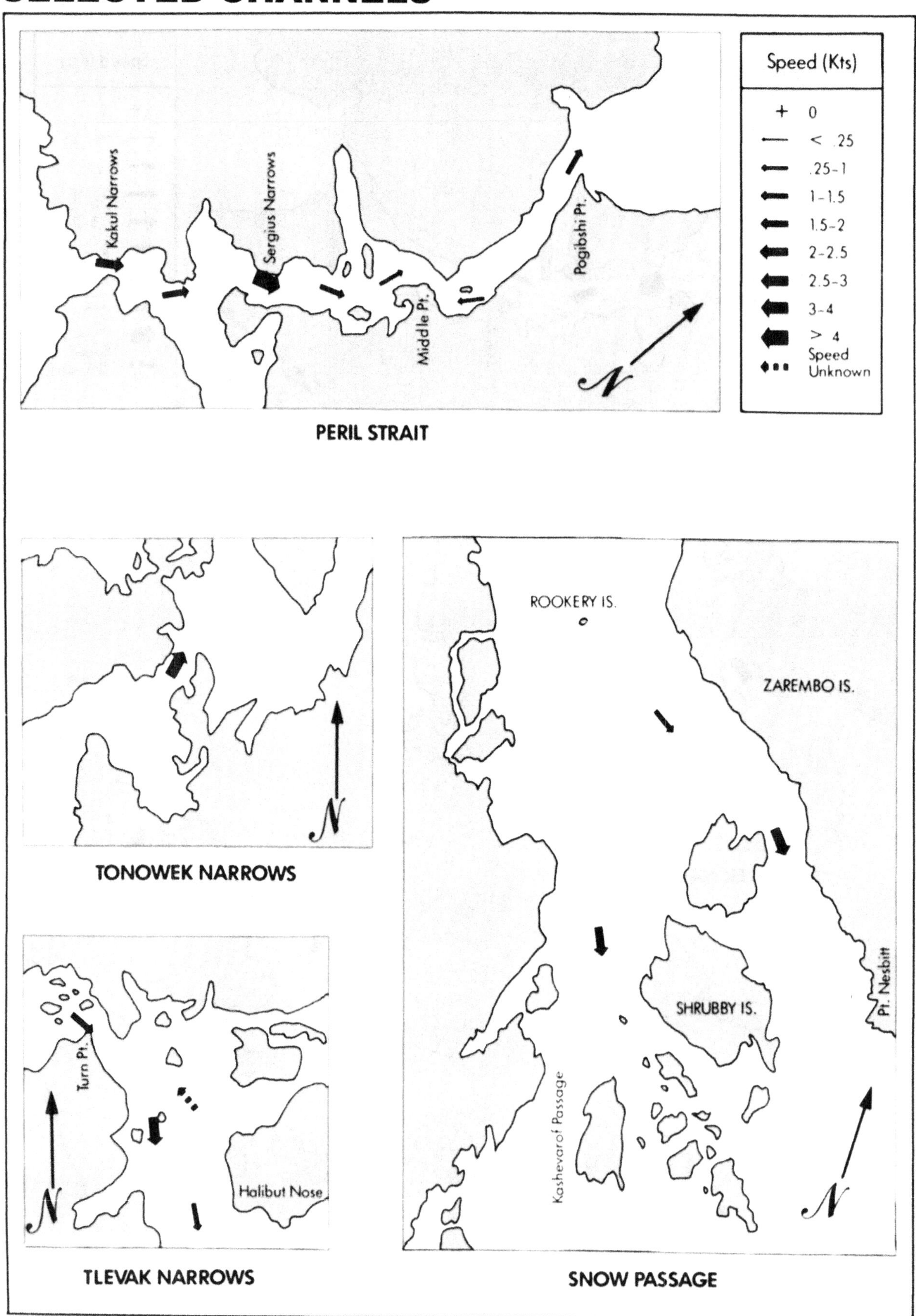
Speed (Kts)
+ 0
< .25
.25-1
1-1.5
1.5-2
2-2.5
2.5-3
3-4
> 4
Speed Unknown
Kakul Narrows
Sergius Narrows
Middle Pt.
Pogibshi Pt.
N
PERIL STRAIT
N
TONOWEK NARROWS
Turn Pt.
Halibut Nose
N
TLEVAK NARROWS
ROOKERY IS.
ZAREMBO IS.
SHRUBBY IS.
Pt. Nesbitt
Kashevarof Passage
N
SNOW PASSAGE

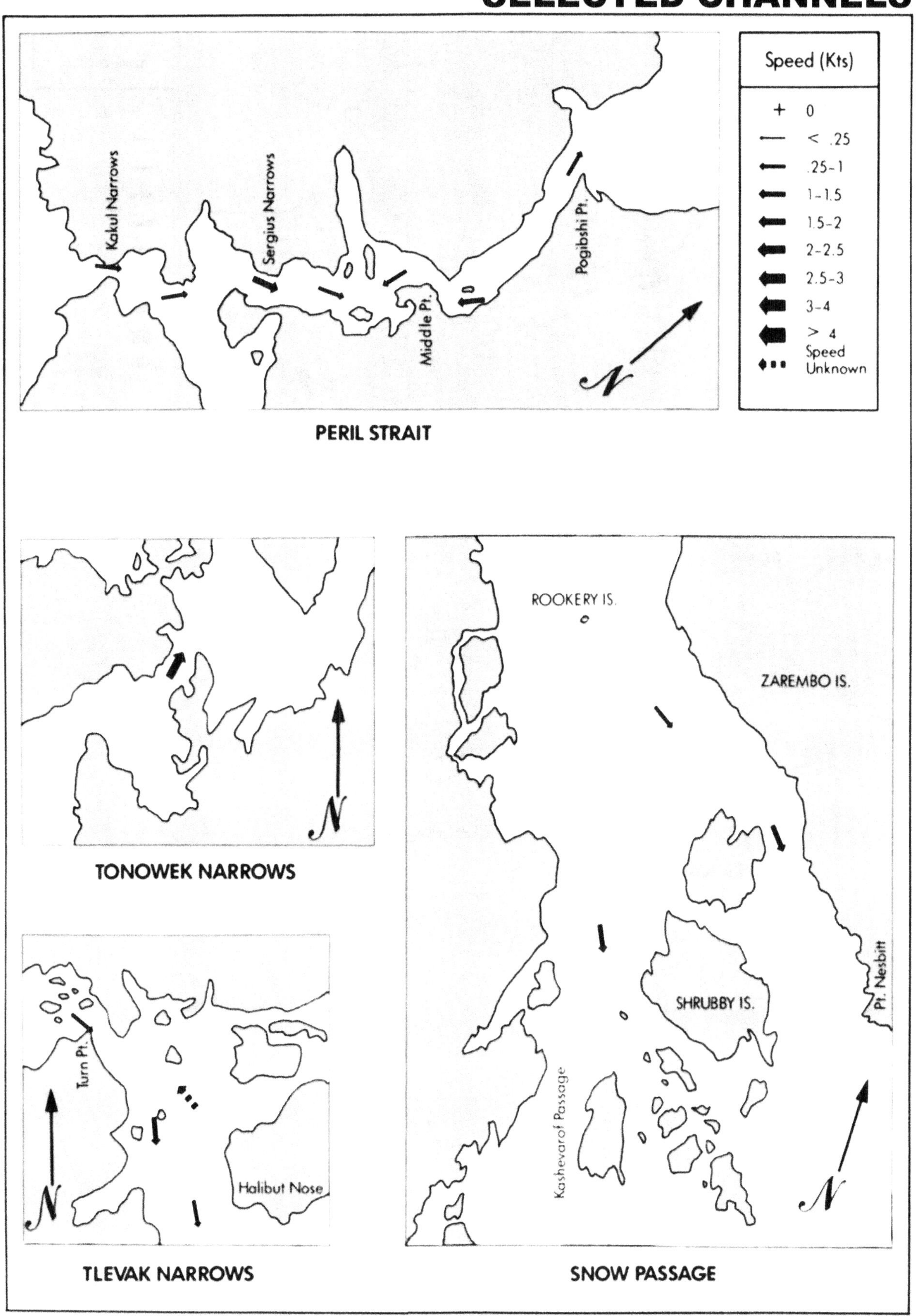
Speed (Kts)
+ 0
< .25
.25-1
1-1.5
1.5-2
2-2.5
2.5-3
3-4
> 4
Speed Unknown
Kakul Narrows
Sergius Narrows
Middle Pt.
Pogibshi Pt.
N
PERIL STRAIT
TONOWEK NARROWS
ROOKERY IS.
ZAREMBO IS.
SHRUBBY IS.
Pt. Nesbitt
Kashevarof Passage
SNOW PASSAGE
Turn Pt.
Halibut Nose
TLEVAK NARROWS

SELECTED CHANNELS

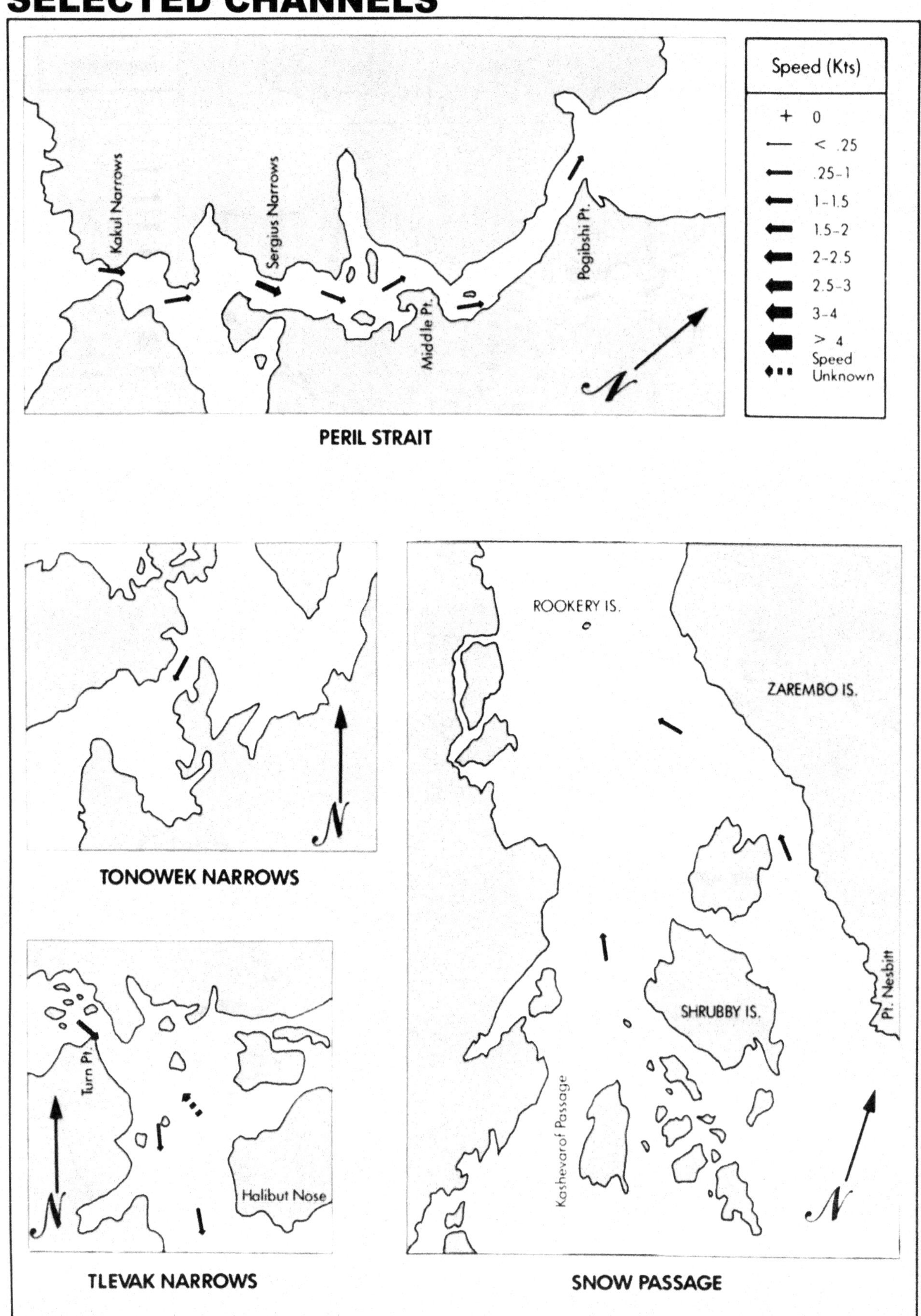
Speed (Kts)
+ 0
< .25
.25-1
1-1.5
1.5-2
2-2.5
2.5-3
3-4
> 4
Speed Unknown
Kakul Narrows
Sergius Narrows
Middle Pt.
Pogibshi Pt.
N
PERIL STRAIT
N
TONOWEK NARROWS
Turn Pt.
Halibut Nose
N
TLEVAK NARROWS
ROOKERY IS.
ZAREMBO IS.
SHRUBBY IS.
Pt. Nesbitt
Kashevarof Passage
N
SNOW PASSAGE

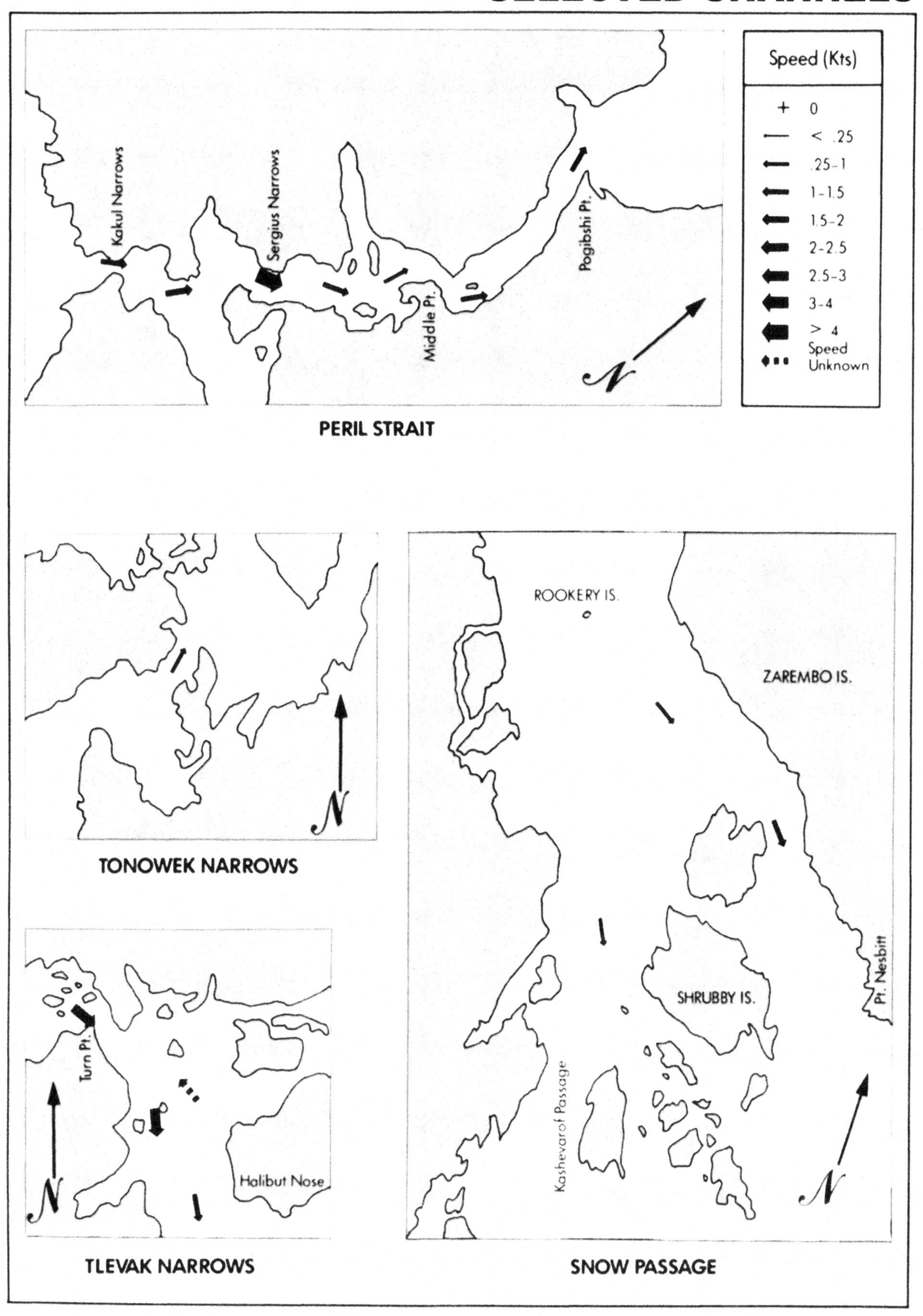
Speed (Kts)
+ 0
< .25
.25-1
1-1.5
1.5-2
2-2.5
2.5-3
3-4
> 4
Speed Unknown
Kakul Narrows
Sergius Narrows
Middle Pt.
Pogibshi Pt.
N
PERIL STRAIT
N
TONOWEK NARROWS
Turn Pt.
Halibut Nose
N
TLEVAK NARROWS
ROOKERY IS.
ZAREMBO IS.
SHRUBBY IS.
Pt. Nesbitt
Kashevarof Passage
N
SNOW PASSAGE

SELECTED CHANNELS

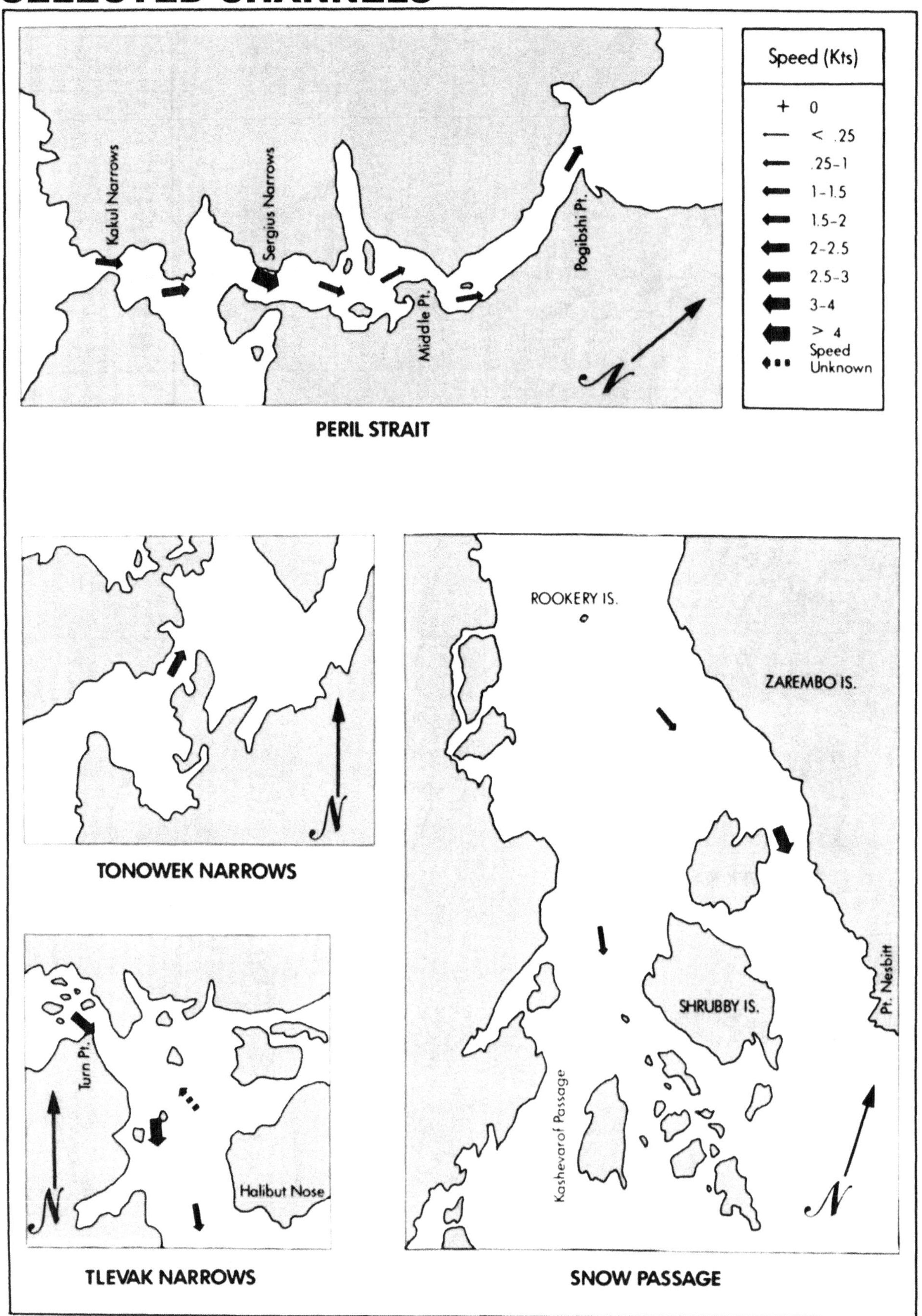
Speed (Kts)
+ 0
< .25
.25-1
1-1.5
1.5-2
2-2.5
2.5-3
3-4
> 4
Speed Unknown
Kakul Narrows
Sergius Narrows
Pogibshi Pt.
Middle Pt.
N
PERIL STRAIT
N
TONOWEK NARROWS
Turn Pt.
Halibut Nose
N
TLEVAK NARROWS
ROOKERY IS.
ZAREMBO IS.
Pt. Nesbitt
SHRUBBY IS.
Kashevarof Passage
N
SNOW PASSAGE

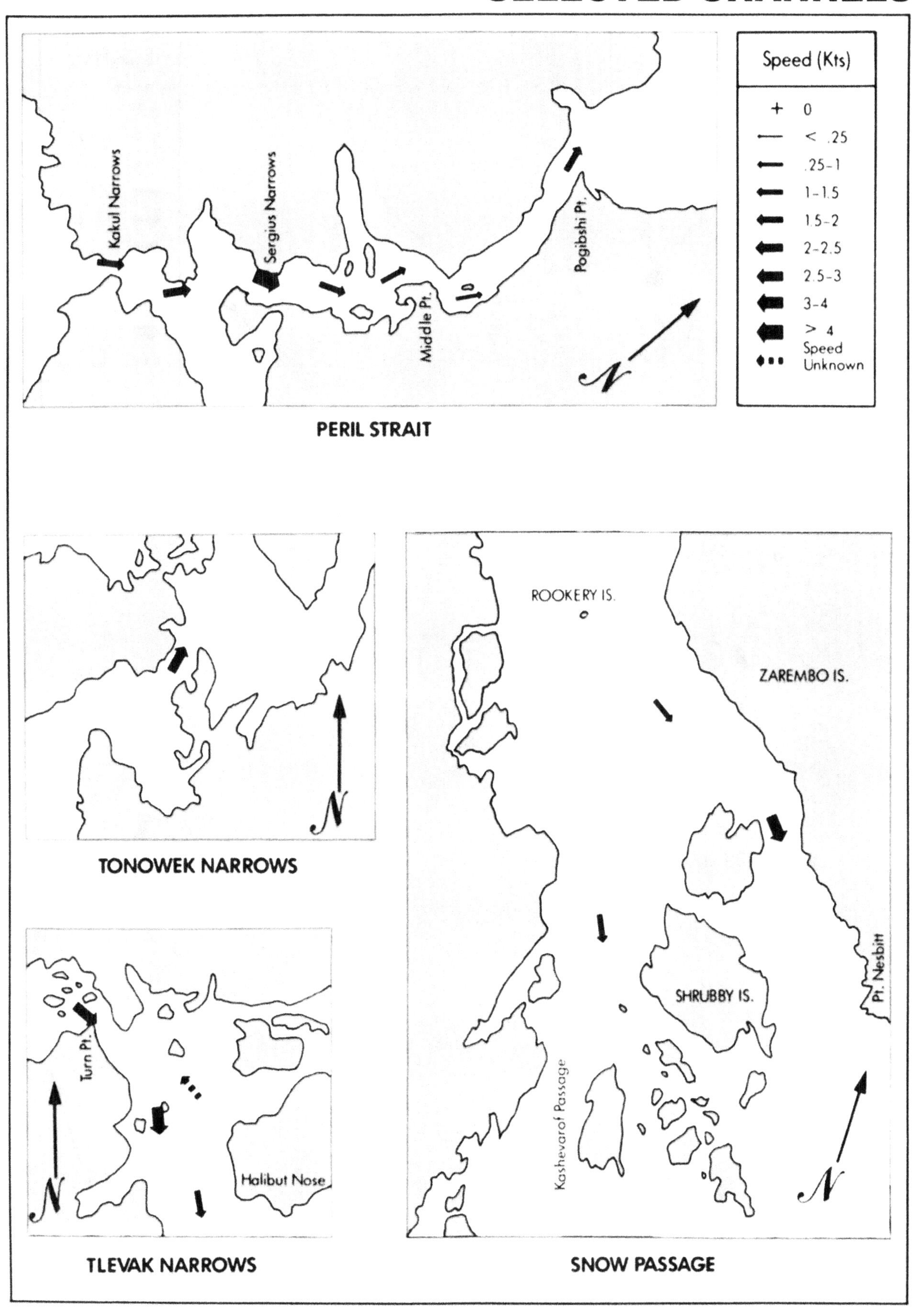
Speed (Kts)
+ 0
< .25
.25-1
1-1.5
1.5-2
2-2.5
2.5-3
3-4
> 4
Speed Unknown
Kakul Narrows
Sergius Narrows
Middle Pt.
Pogibshi Pt.
N
PERIL STRAIT
N
TONOWEK NARROWS
Turn Pt.
Halibut Nose
N
TLEVAK NARROWS
ROOKERY IS.
ZAREMBO IS.
SHRUBBY IS.
Pt. Nesbitt
Kashevarof Passage
N
SNOW PASSAGE

SELECTED CHANNELS

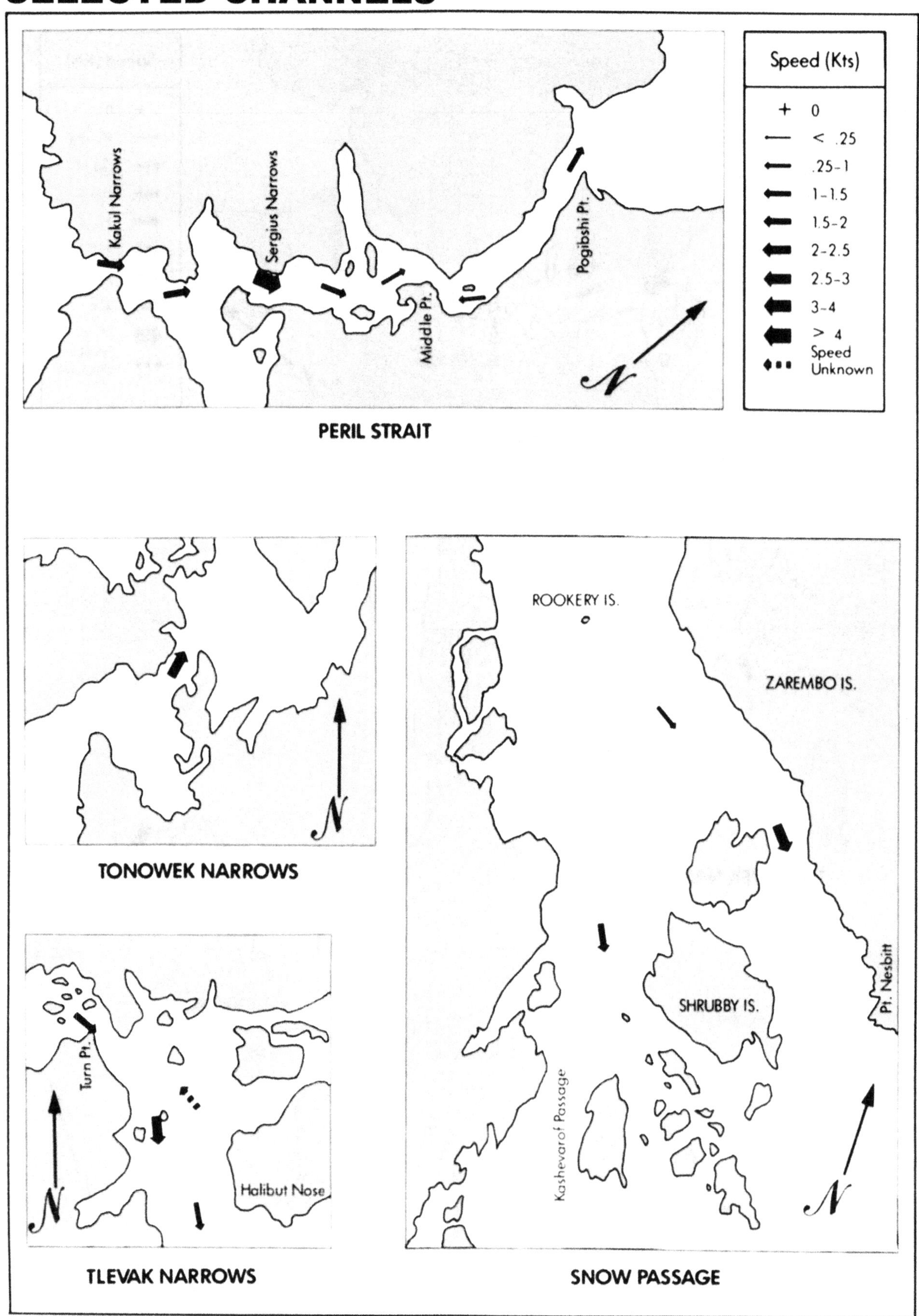
Speed (Kts)
+ 0
< .25
.25-1
1-1.5
1.5-2
2-2.5
2.5-3
3-4
> 4
Speed Unknown
Kakul Narrows
Sergius Narrows
Middle Pt.
Pogibshi Pt.
PERIL STRAIT
TONOWEK NARROWS
Turn Pt.
Halibut Nose
TLEVAK NARROWS
ROOKERY IS.
ZAREMBO IS.
SHRUBBY IS.
Pt. Nesbitt
Kashevarof Passage
SNOW PASSAGE

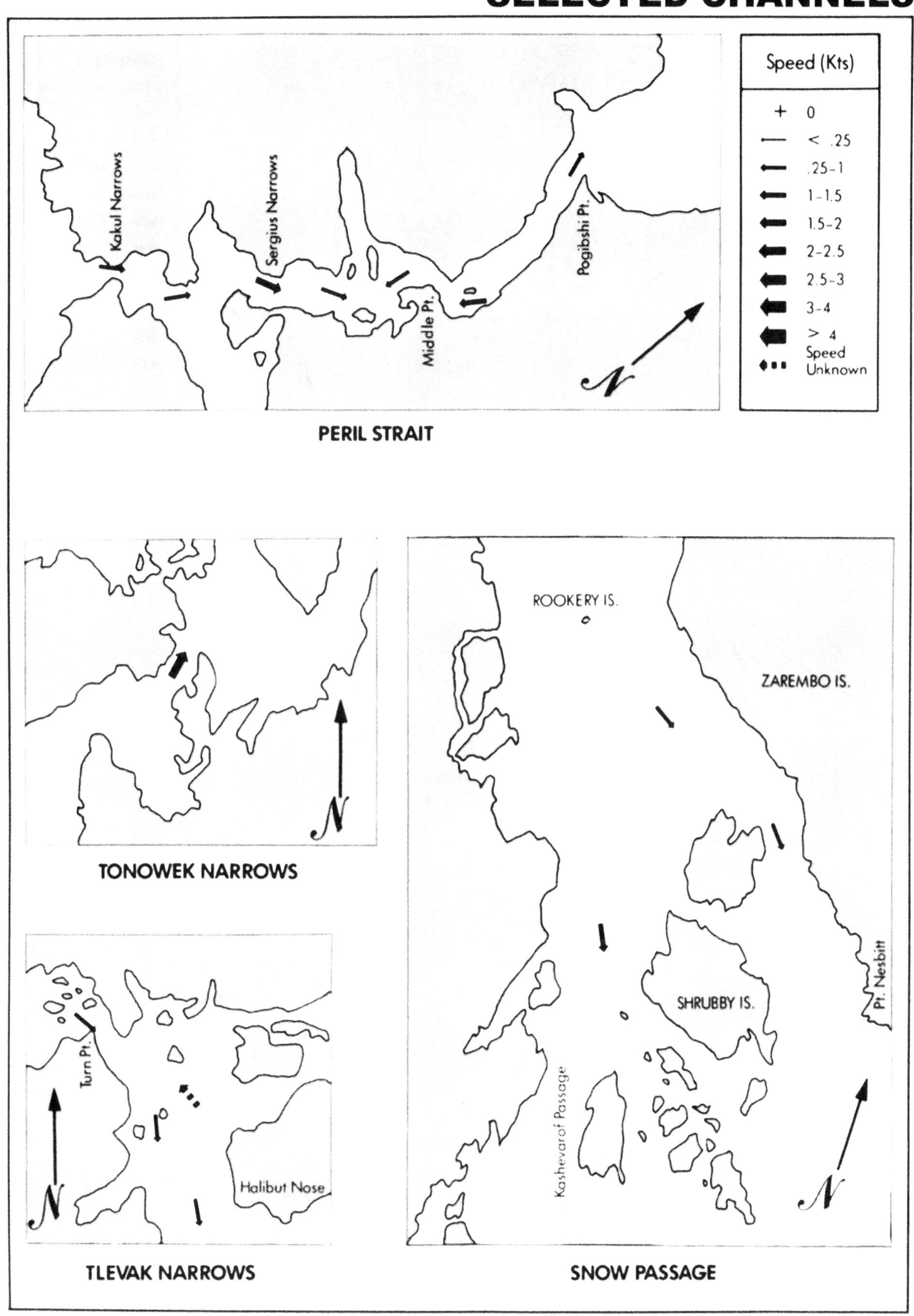
Speed (Kts)
0
< .25
.25-1
1-1.5
1.5-2
2-2.5
2.5-3
3-4
> 4
Speed Unknown
Kakul Narrows
Sergius Narrows
Middle Pt.
Pogibshi Pt.
PERIL STRAIT
TONOWEK NARROWS
Turn Pt.
Halibut Nose
TLEVAK NARROWS
ROOKERY IS.
ZAREMBO IS.
SHRUBBY IS.
Pt. Nesbitt
Kashevarof Passage
SNOW PASSAGE

SELECTED CHANNELS

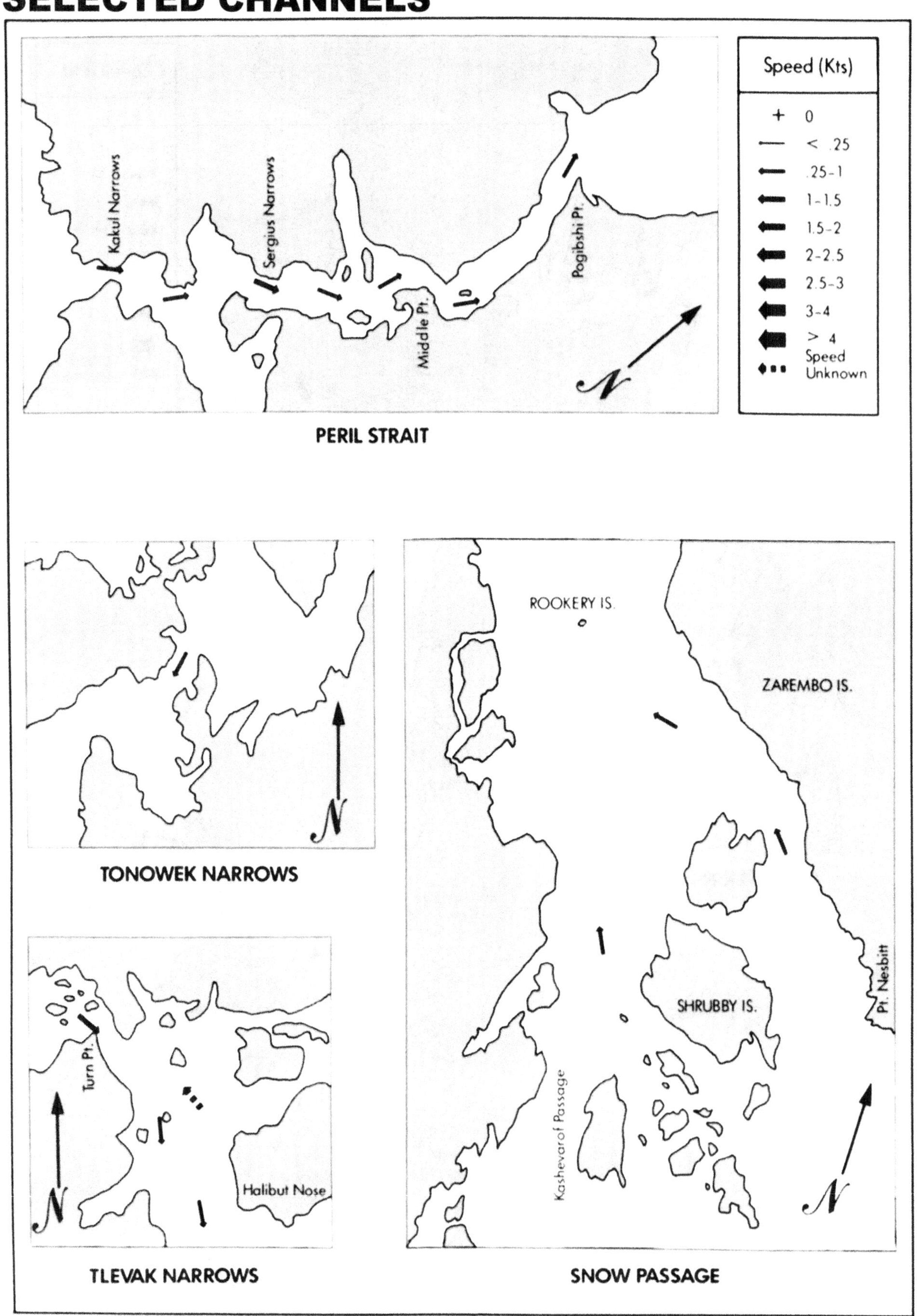

Speed (Kts)
0
< .25
.25-1
1-1.5
1.5-2
2-2.5
2.5-3
3-4
> 4
Speed Unknown
Kakul Narrows
Sergius Narrows
Middle Pt.
Pogibshi Pt.
N
PERIL STRAIT
N
TONOWEK NARROWS
Turn Pt.
Halibut Nose
N
TLEVAK NARROWS
ROOKERY IS.
ZAREMBO IS.
SHRUBBY IS.
Pt. Nesbitt
Kashevarof Passage
N
SNOW PASSAGE

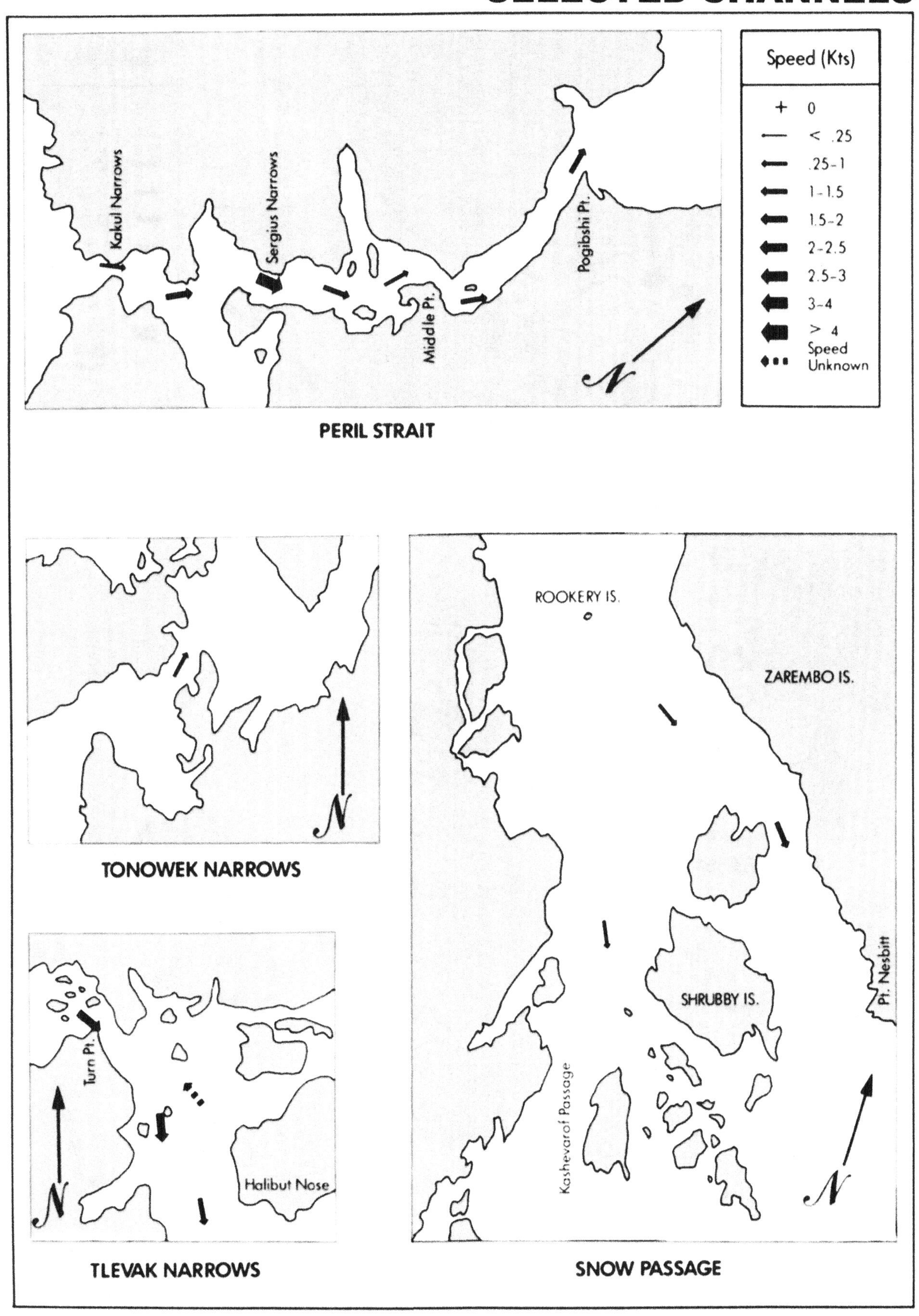
Speed (Kts)
+ 0
< .25
.25-1
1-1.5
1.5-2
2-2.5
2.5-3
3-4
> 4
Speed Unknown
Kakul Narrows
Sergius Narrows
Middle Pt.
Pogibshi Pt.
N
PERIL STRAIT
N
TONOWEK NARROWS
Turn Pt.
Halibut Nose
N
TLEVAK NARROWS
ROOKERY IS.
ZAREMBO IS.
SHRUBBY IS.
Pt. Nesbitt
Kashevarof Passage
N
SNOW PASSAGE

SELECTED CHANNELS

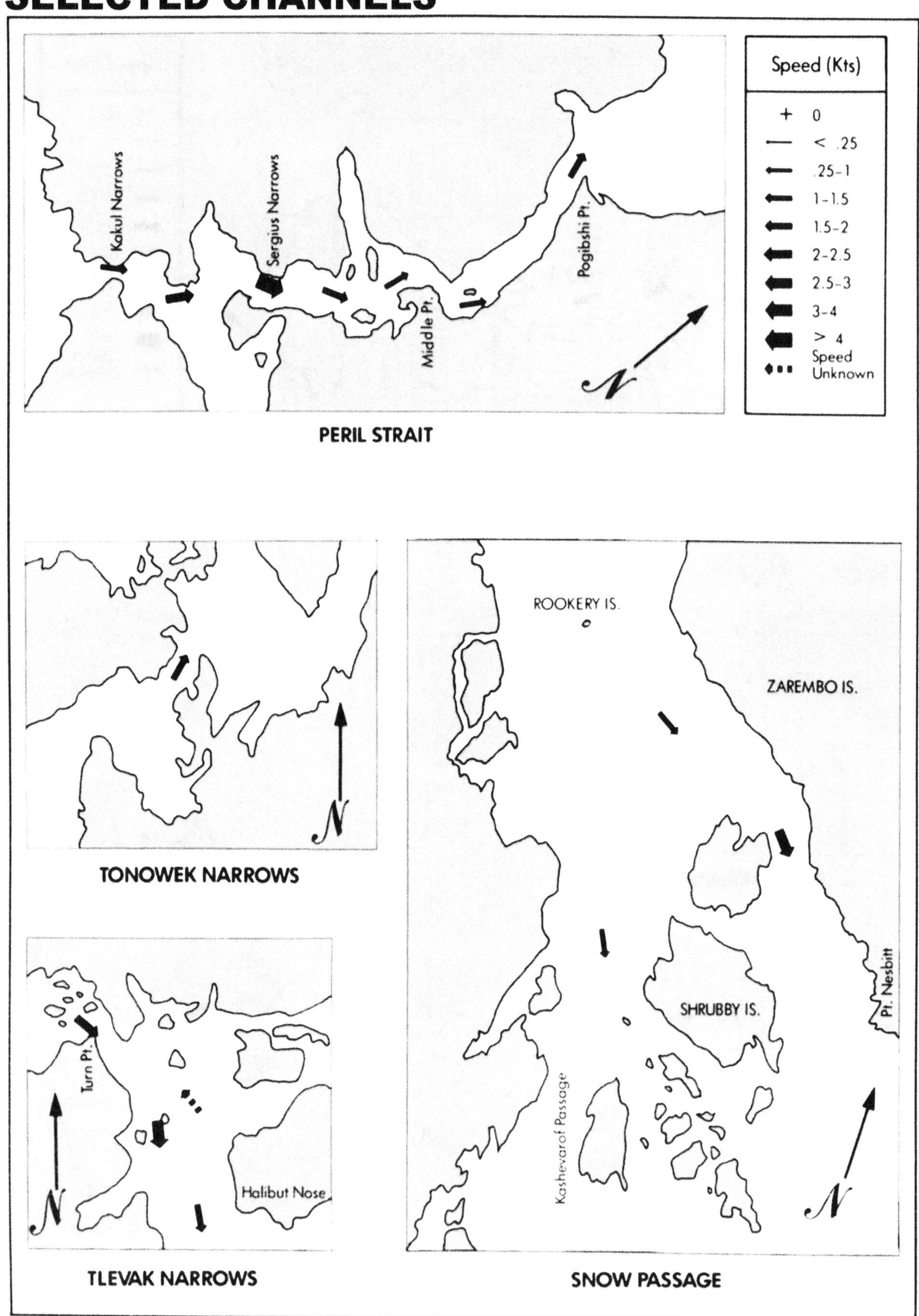
Speed (Kts)
+ 0
< .25
.25-1
1-1.5
1.5-2
2-2.5
2.5-3
3-4
> 4
Speed Unknown
Kakul Narrows
Sergius Narrows
Middle Pt.
Pogibshi Pt.
N
PERIL STRAIT
N
TONOWEK NARROWS
Turn Pt.
Halibut Nose
N
TLEVAK NARROWS
ROOKERY IS.
ZAREMBO IS.
SHRUBBY IS.
Pt. Nesbitt
Kashevarof Passage
N
SNOW PASSAGE

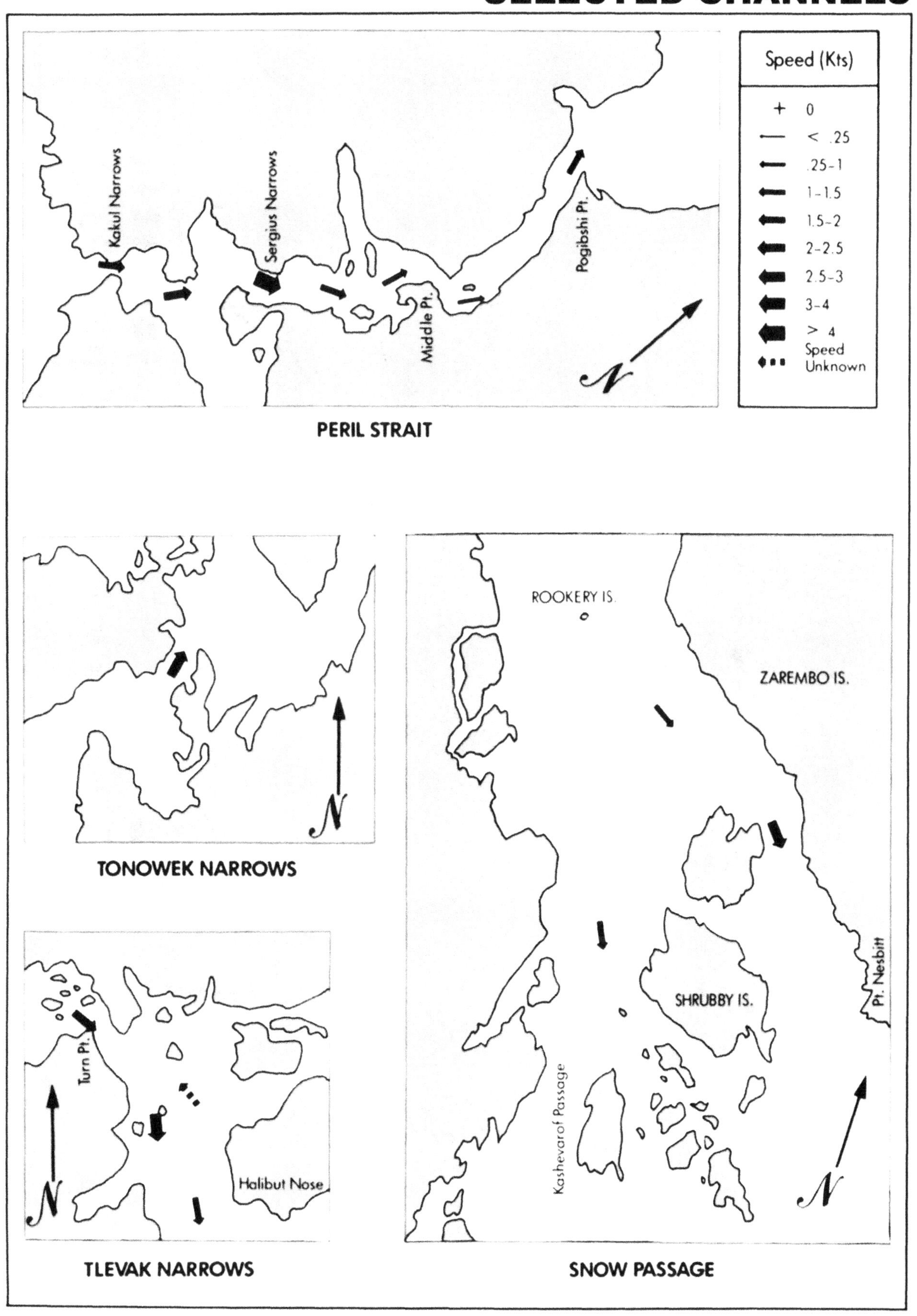
Speed (Kts)
+ 0
< .25
.25-1
1-1.5
1.5-2
2-2.5
2.5-3
3-4
> 4
Speed Unknown
Kakul Narrows
Sergius Narrows
Middle Pt.
Pogibshi Pt.
N
PERIL STRAIT
TONOWEK NARROWS
ROOKERY IS.
ZAREMBO IS.
SHRUBBY IS.
Pt. Nesbitt
Kashevarof Passage
SNOW PASSAGE
Turn Pt.
Halibut Nose
TLEVAK NARROWS

SELECTED CHANNELS

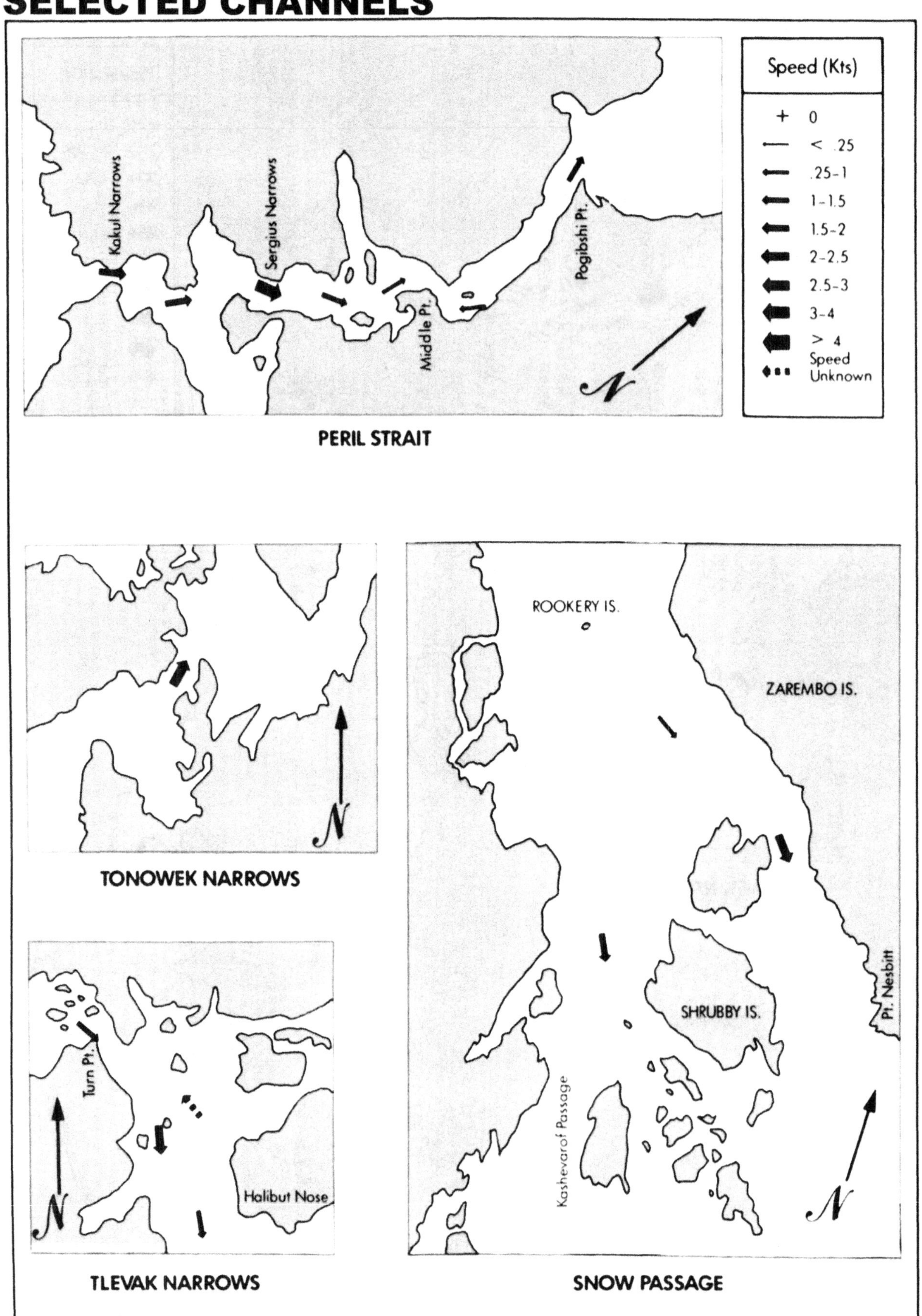

Speed (Kts)
0
< .25
.25-1
1-1.5
1.5-2
2-2.5
2.5-3
3-4
> 4
Speed Unknown
Kakul Narrows
Sergius Narrows
Middle Pt.
Pogibshi Pt.
PERIL STRAIT
TONOWEK NARROWS
ROOKERY IS.
ZAREMBO IS.
SHRUBBY IS.
Pt. Nesbitt
Kashevarof Passage
SNOW PASSAGE
Turn Pt.
Halibut Nose
TLEVAK NARROWS

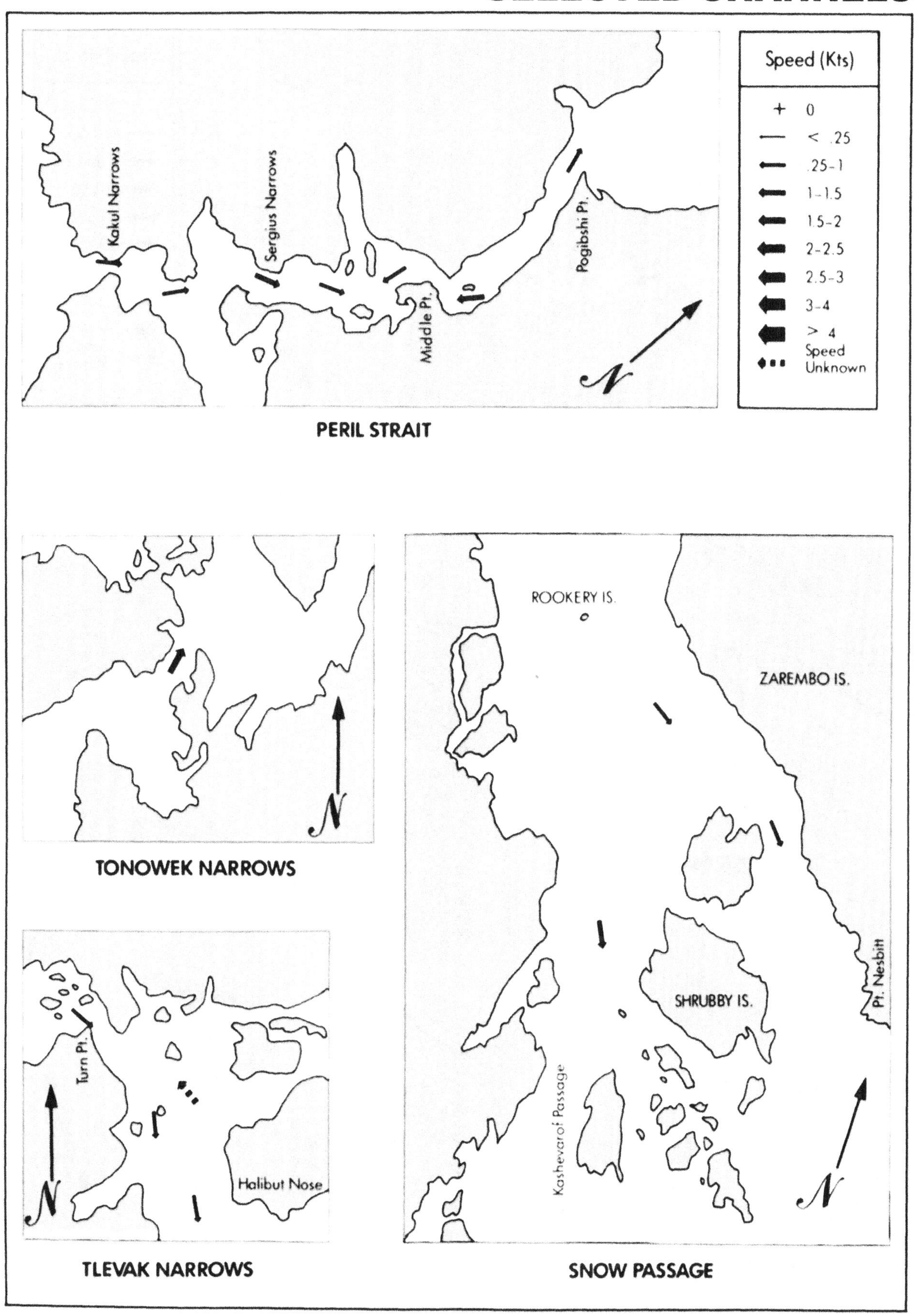
Speed (Kts)
+ 0
< .25
.25-1
1-1.5
1.5-2
2-2.5
2.5-3
3-4
> 4
Speed Unknown
Kakul Narrows
Sergius Narrows
Middle Pt.
Pogibshi Pt.
PERIL STRAIT
TONOWEK NARROWS
Turn Pt.
Halibut Nose
TLEVAK NARROWS
ROOKERY IS.
ZAREMBO IS.
SHRUBBY IS.
Pt. Nesbitt
Kashevarof Passage
SNOW PASSAGE

SELECTED CHANNELS

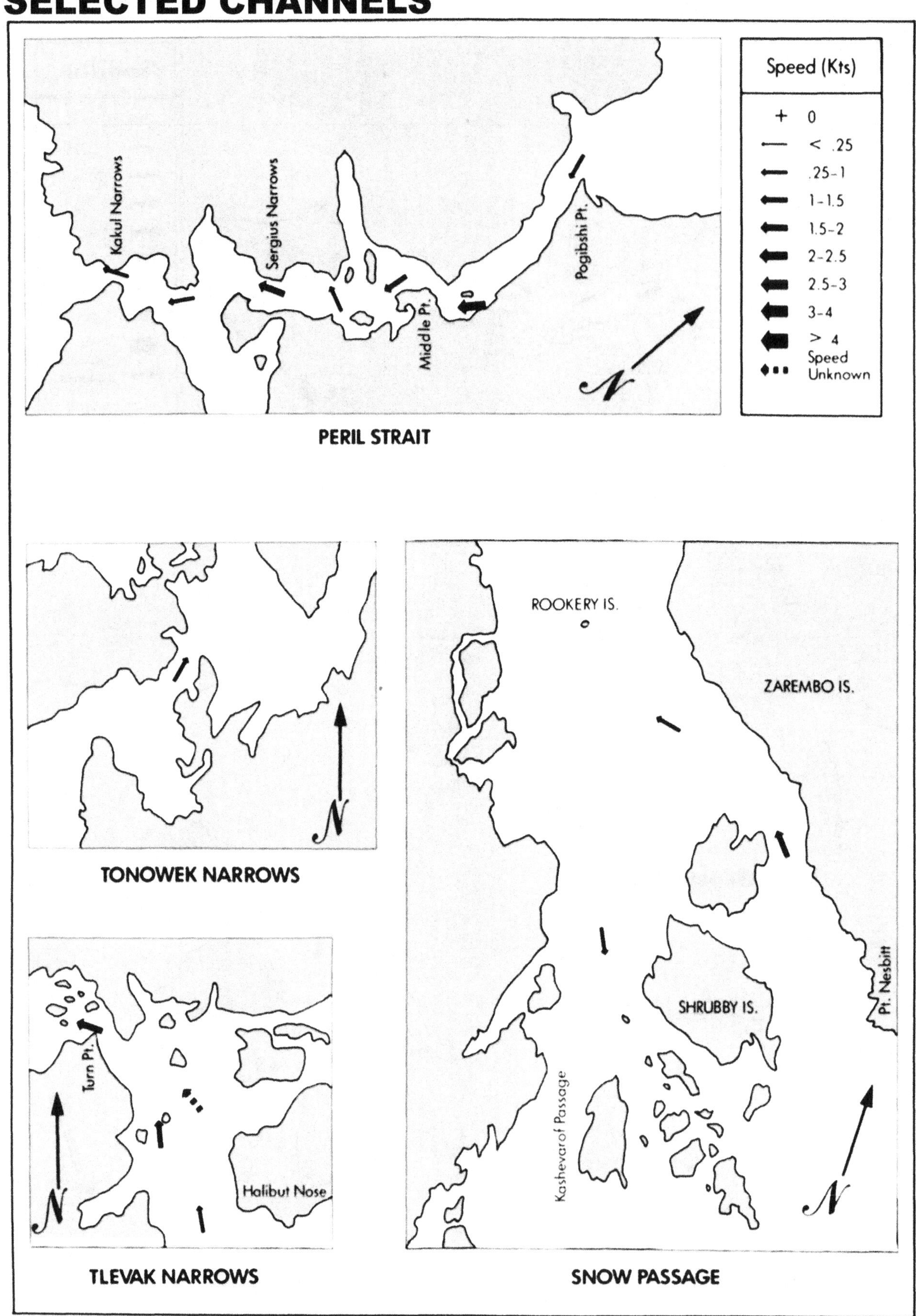
Speed (Kts)
+ 0
< .25
.25-1
1-1.5
1.5-2
2-2.5
2.5-3
3-4
> 4
Speed Unknown
Kakul Narrows
Sergius Narrows
Middle Pt.
Pogibshi Pt.
N
PERIL STRAIT
N
TONOWEK NARROWS
Turn Pt.
Halibut Nose
N
TLEVAK NARROWS
ROOKERY IS.
ZAREMBO IS.
SHRUBBY IS.
Pt. Nesbitt
Kashevarof Passage
N
SNOW PASSAGE

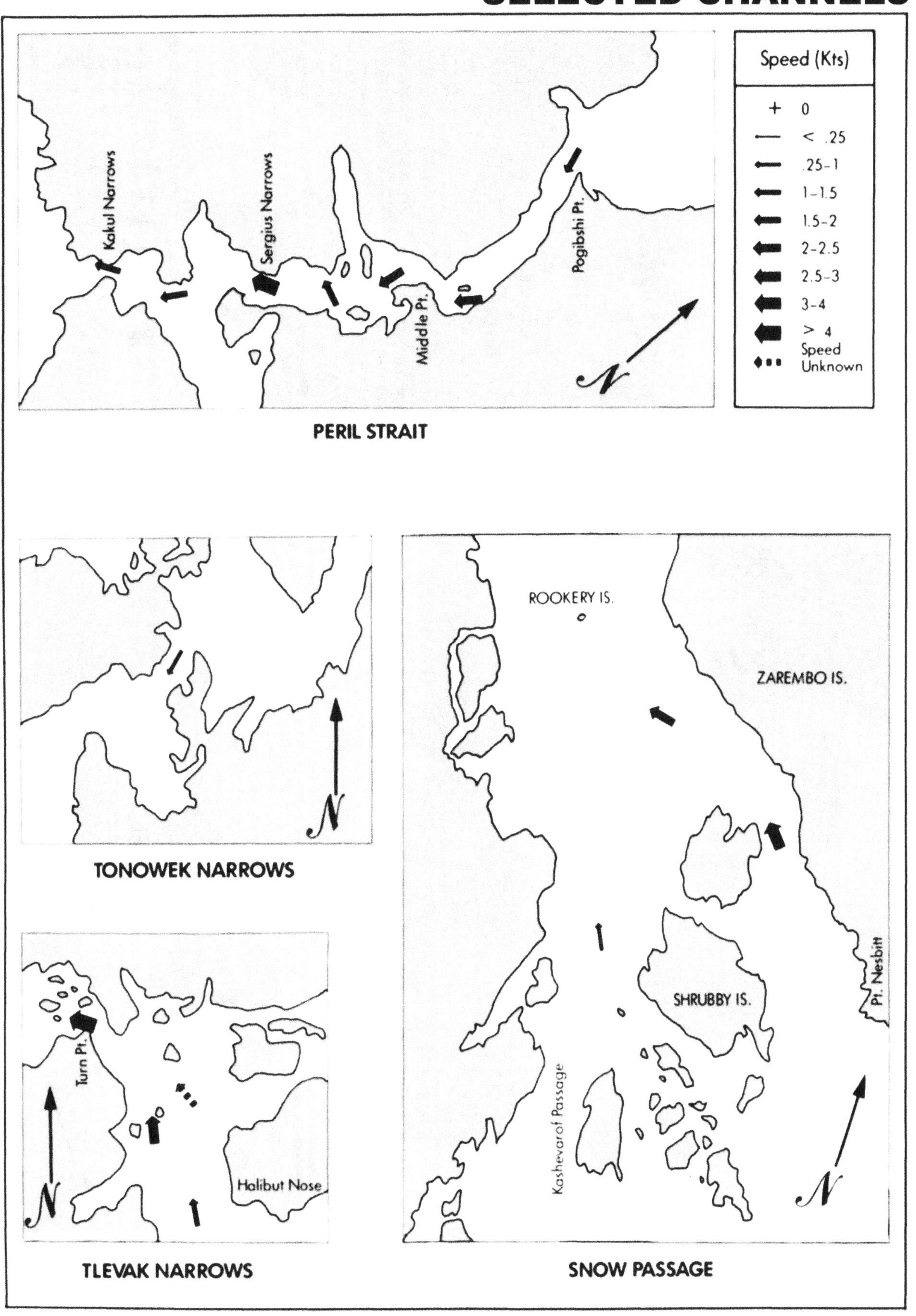
Speed (Kts)
+ 0
< .25
.25-1
1-1.5
1.5-2
2-2.5
2.5-3
3-4
> 4
Speed Unknown
Kakul Narrows
Sergius Narrows
Middle Pt.
Pogibshi Pt.
N
PERIL STRAIT
N
TONOWEK NARROWS
Turn Pt.
Halibut Nose
N
TLEVAK NARROWS
ROOKERY IS.
ZAREMBO IS.
SHRUBBY IS.
Pt. Nesbitt
Kashevarof Passage
N
SNOW PASSAGE

SELECTED CHANNELS

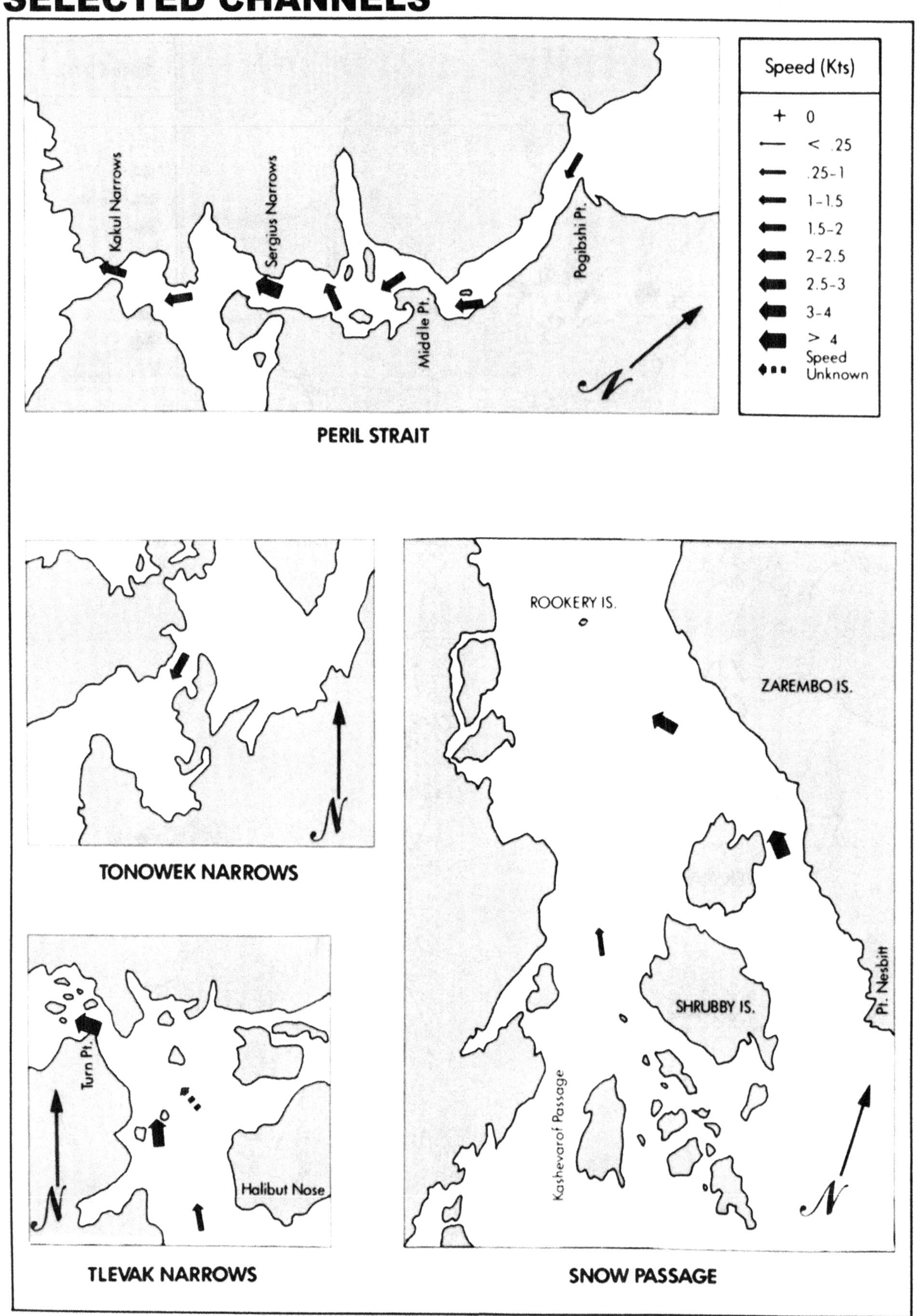

Speed (Kts)
0
< .25
.25-1
1-1.5
1.5-2
2-2.5
2.5-3
3-4
> 4
Speed Unknown
Kakul Narrows
Sergius Narrows
Pogibshi Pt.
Middle Pt.
N
PERIL STRAIT
TONOWEK NARROWS
ROOKERY IS.
ZAREMBO IS.
SHRUBBY IS.
Pt. Nesbitt
Kashevarof Passage
SNOW PASSAGE
Turn Pt.
Halibut Nose
TLEVAK NARROWS

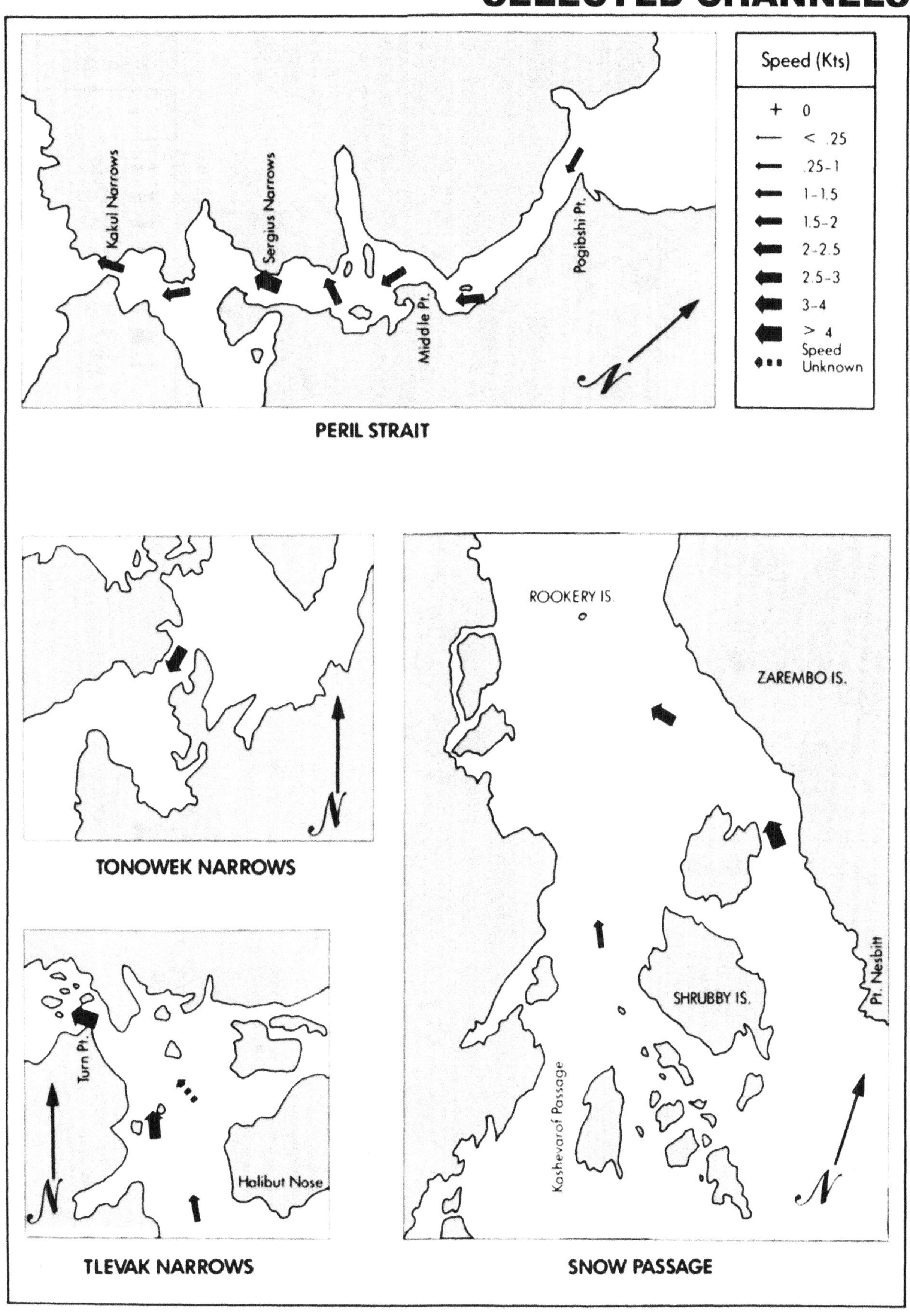
Speed (Kts)
+ 0
< .25
.25-1
1-1.5
1.5-2
2-2.5
2.5-3
3-4
> 4
Speed Unknown
Kakul Narrows
Sergius Narrows
Middle Pt.
Pogibshi Pt.
N
PERIL STRAIT
N
TONOWEK NARROWS
Turn Pt.
Halibut Nose
N
TLEVAK NARROWS
ROOKERY IS.
ZAREMBO IS.
SHRUBBY IS.
Pt. Nesbitt
Kashevarof Passage
N
SNOW PASSAGE

SELECTED CHANNELS

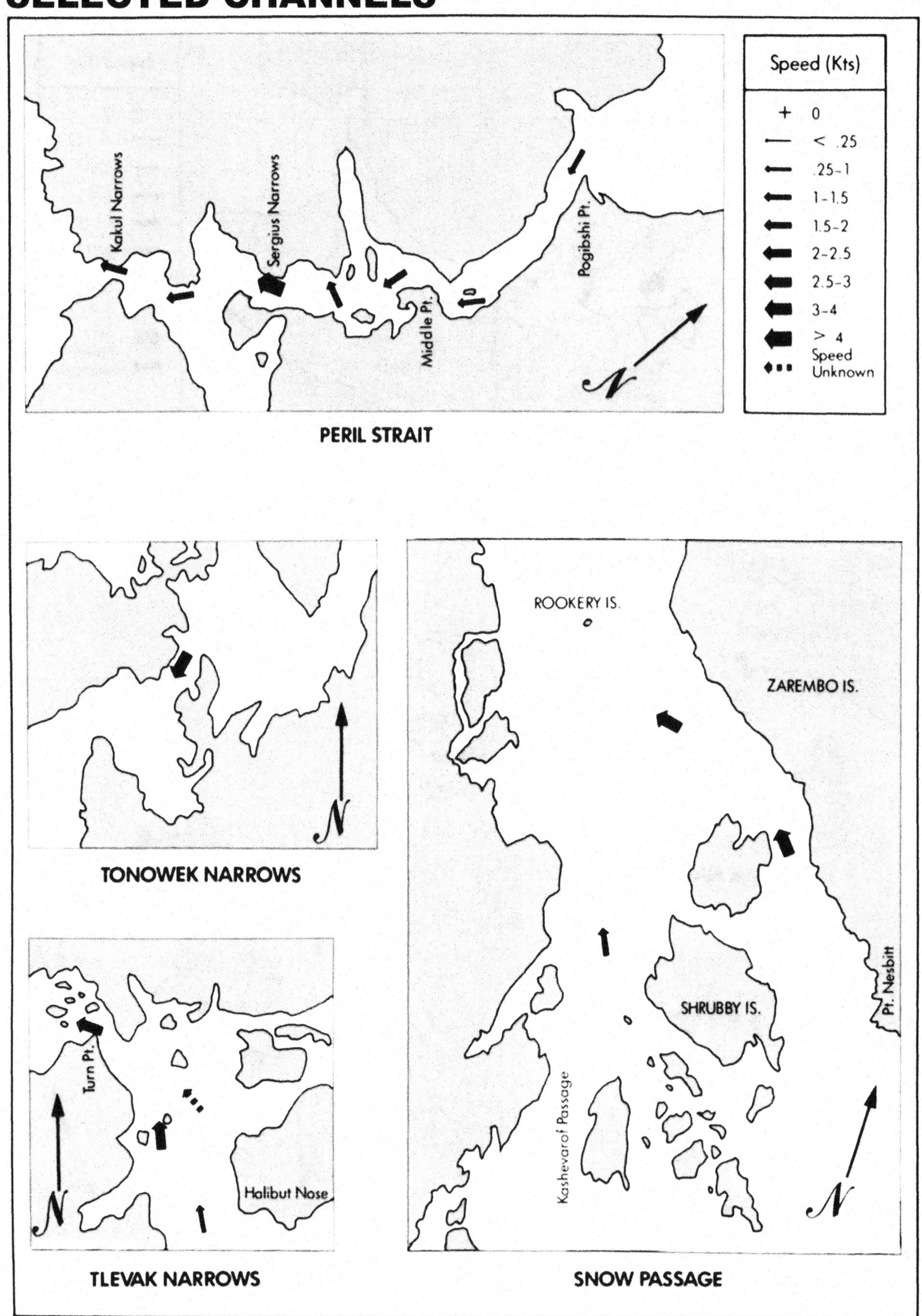

Speed (Kts)
0
< .25
.25-1
1-1.5
1.5-2
2-2.5
2.5-3
3-4
> 4
Speed Unknown
Kakul Narrows
Sergius Narrows
Middle Pt.
Pogibshi Pt.
N
PERIL STRAIT
N
TONOWEK NARROWS
Turn Pt.
Halibut Nose
N
TLEVAK NARROWS
ROOKERY IS.
ZAREMBO IS.
SHRUBBY IS.
Pt. Nesbitt
Kashevarof Passage
N
SNOW PASSAGE

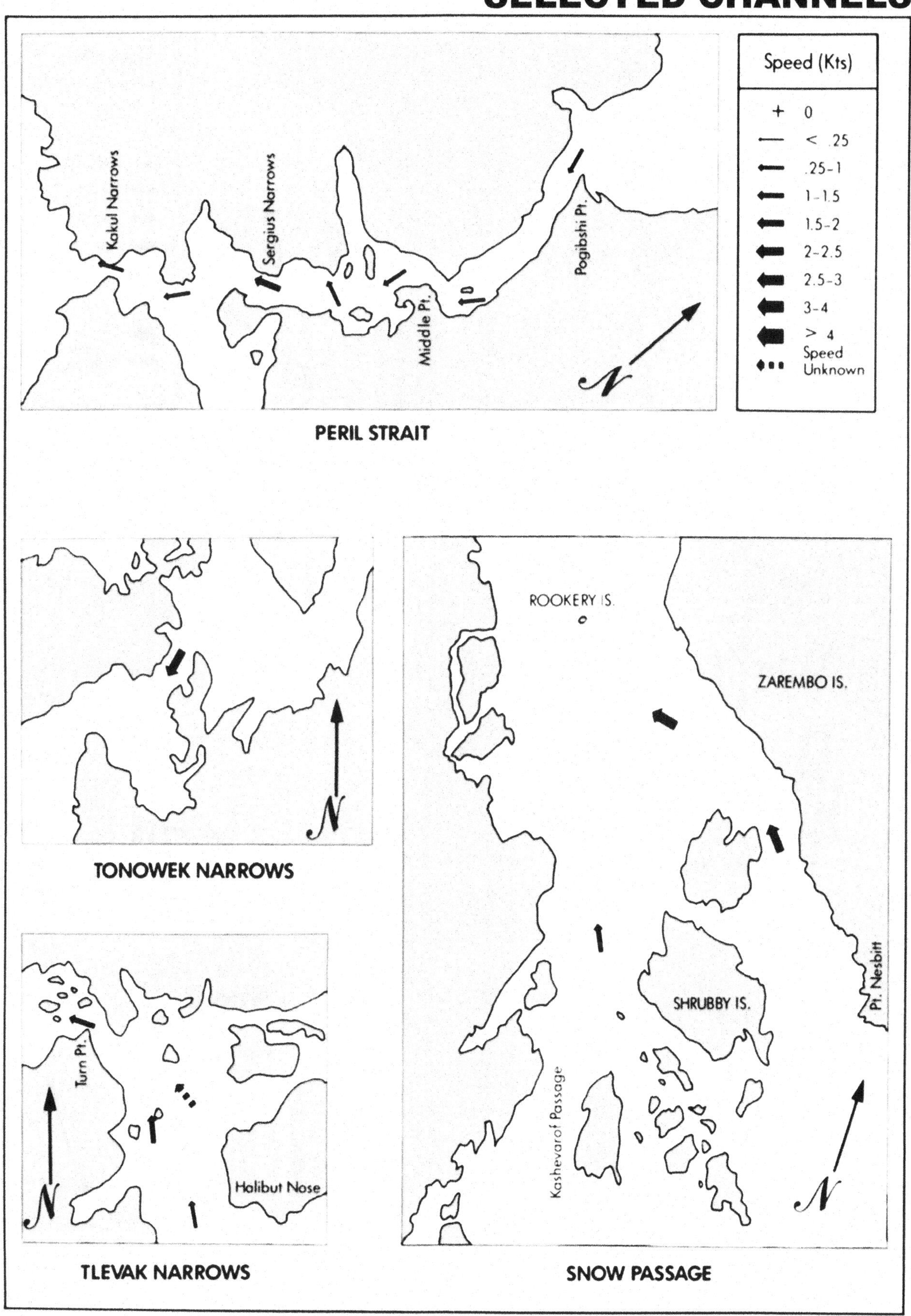
Speed (Kts)
+ 0
< .25
.25-1
1-1.5
1.5-2
2-2.5
2.5-3
3-4
> 4
Speed Unknown
Kakul Narrows
Sergius Narrows
Middle Pt.
Pogibshi Pt.
PERIL STRAIT
TONOWEK NARROWS
Turn Pt.
Halibut Nose
TLEVAK NARROWS
ROOKERY IS.
ZAREMBO IS.
SHRUBBY IS.
Pt. Nesbitt
Kashevarof Passage
SNOW PASSAGE

SELECTED CHANNELS

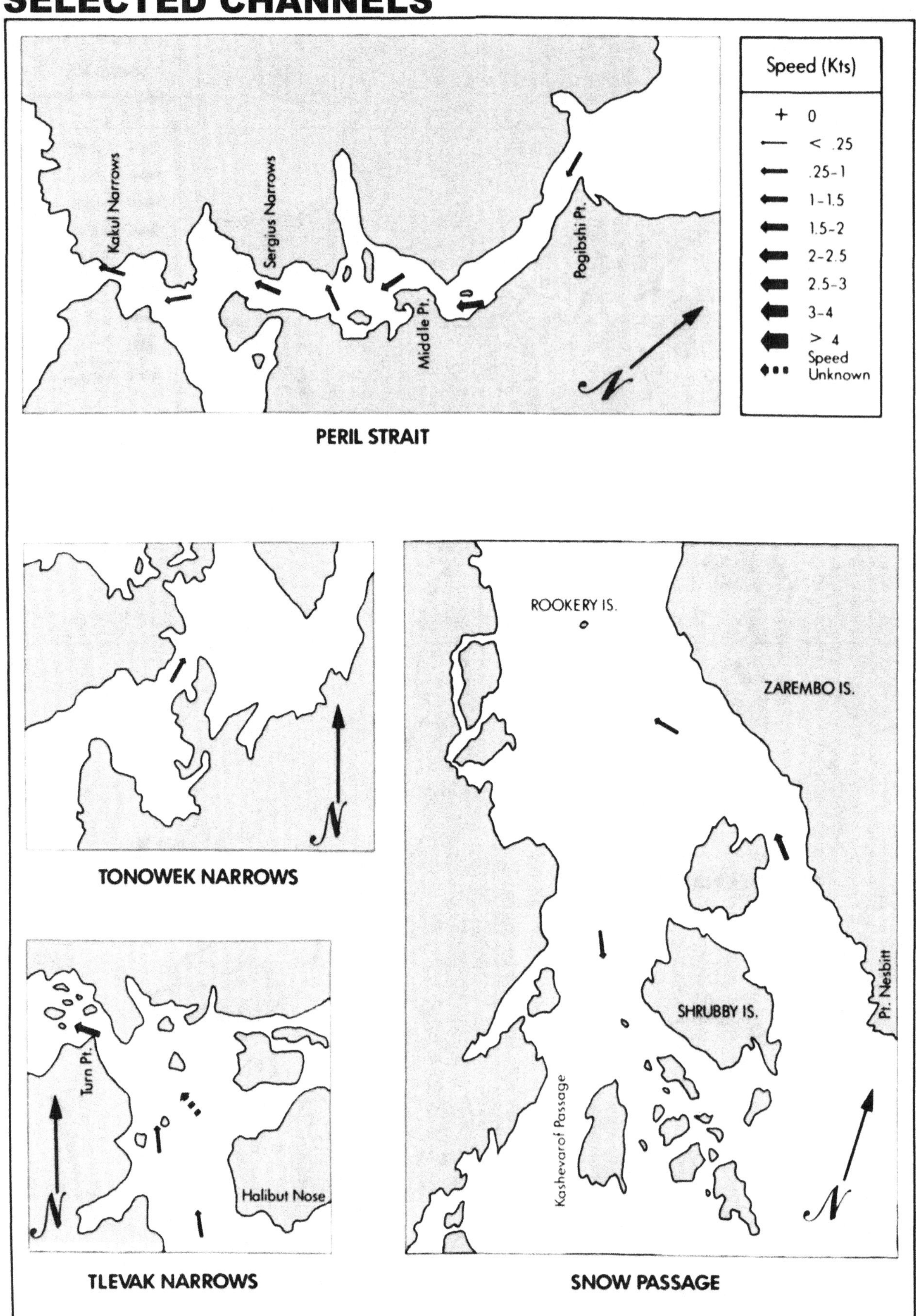
Speed (Kts)
+ 0
< .25
.25-1
1-1.5
1.5-2
2-2.5
2.5-3
3-4
> 4
Speed Unknown
Kakul Narrows
Sergius Narrows
Middle Pt.
Pogibshi Pt.
N
PERIL STRAIT
N
TONOWEK NARROWS
Turn Pt.
Halibut Nose
N
TLEVAK NARROWS
ROOKERY IS.
ZAREMBO IS.
SHRUBBY IS.
Pt. Nesbitt
Kashevarof Passage
N
SNOW PASSAGE

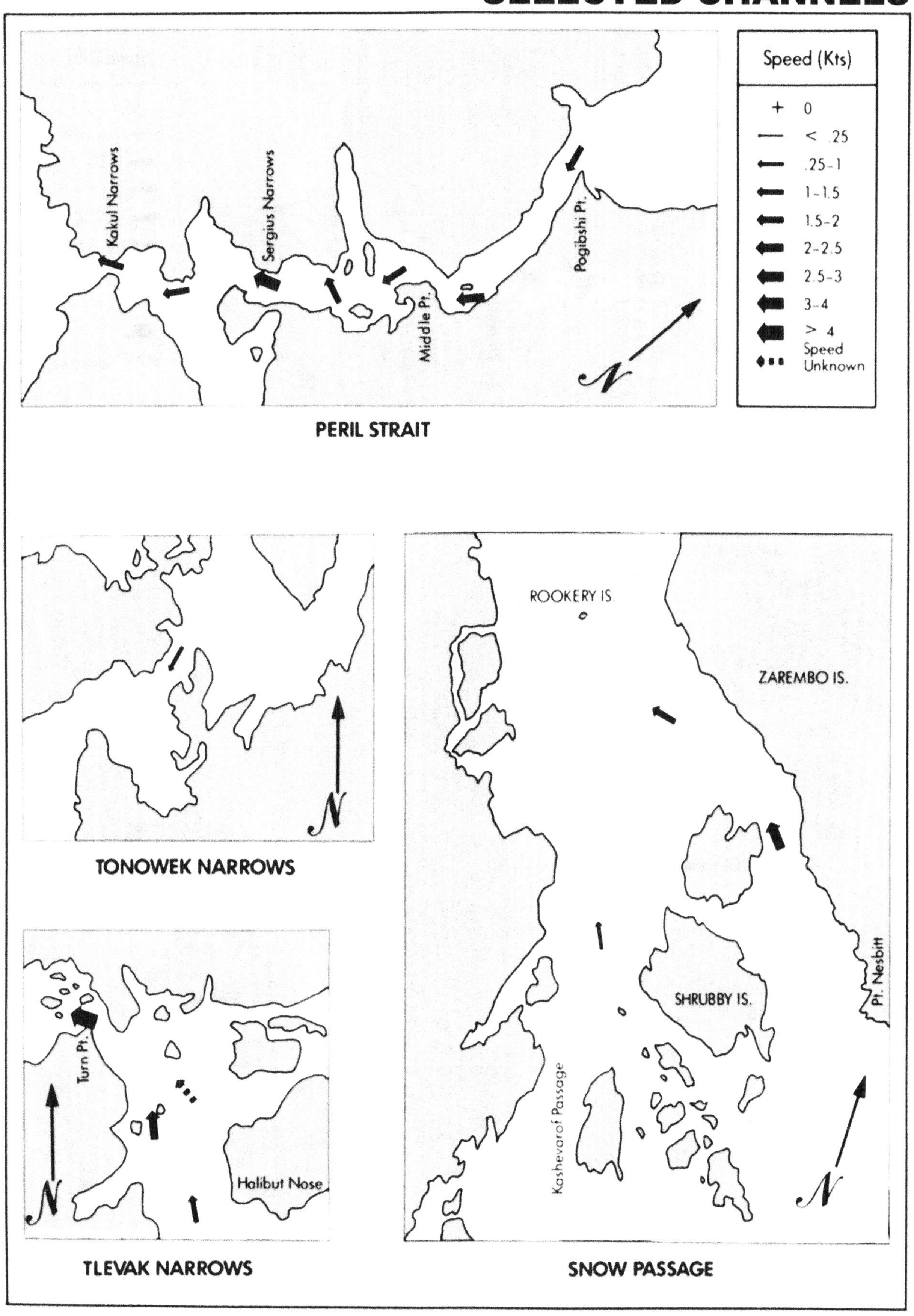
Speed (Kts)
+ 0
< .25
.25-1
1-1.5
1.5-2
2-2.5
2.5-3
3-4
> 4
Speed Unknown
Kakul Narrows
Sergius Narrows
Middle Pt.
Pogibshi Pt.
N
PERIL STRAIT
TONOWEK NARROWS
Turn Pt.
Halibut Nose
TLEVAK NARROWS
ROOKERY IS.
ZAREMBO IS.
SHRUBBY IS.
Pt. Nesbitt
Kashevarof Passage
SNOW PASSAGE

SELECTED CHANNELS

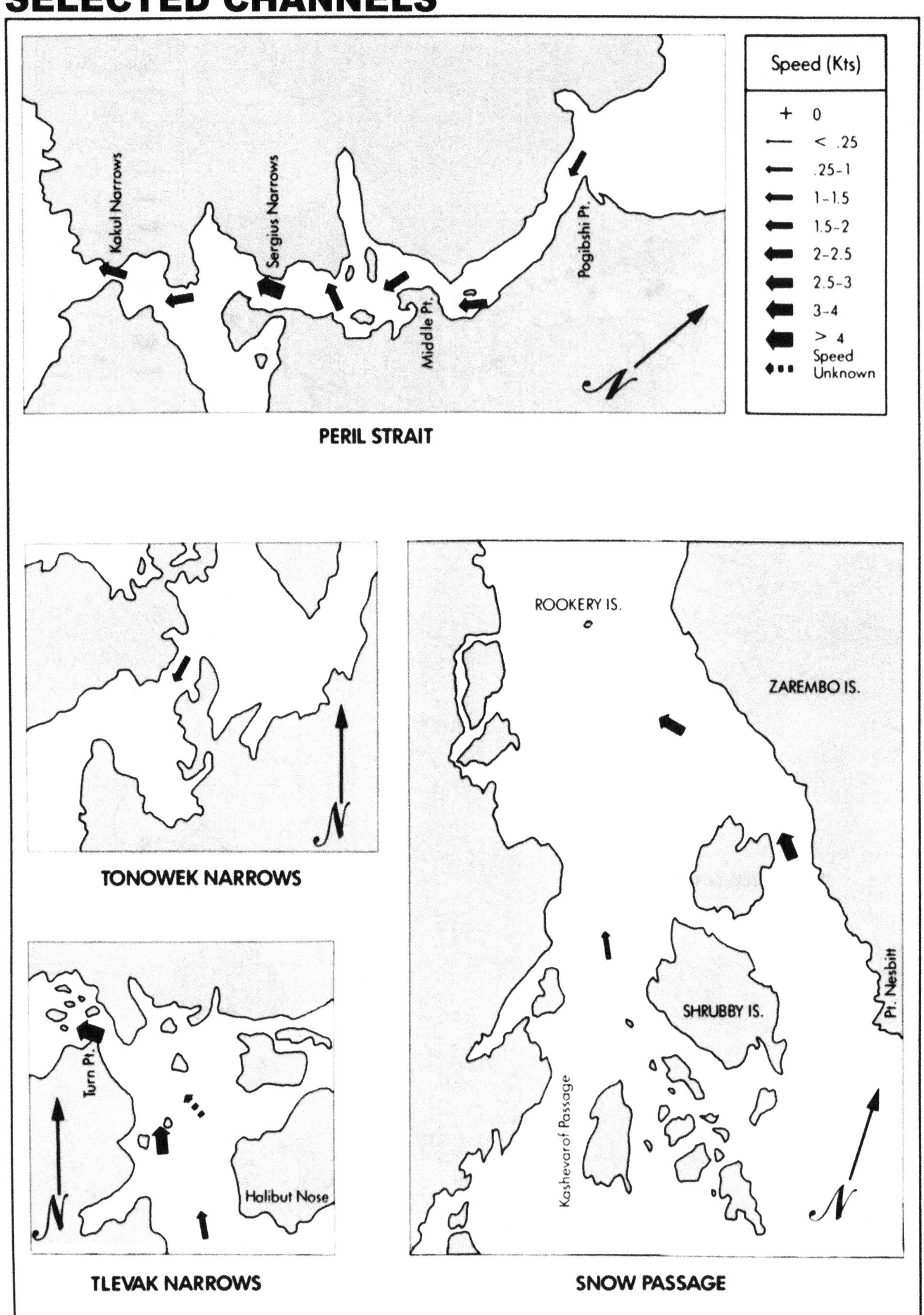

Speed (Kts)
+ 0
< .25
.25-1
1-1.5
1.5-2
2-2.5
2.5-3
3-4
> 4
Speed Unknown
Kakul Narrows
Sergius Narrows
Middle Pt.
Pogibshi Pt.
N
PERIL STRAIT
N
TONOWEK NARROWS
Turn Pt.
Halibut Nose
N
TLEVAK NARROWS
ROOKERY IS.
ZAREMBO IS.
SHRUBBY IS.
Pt. Nesbitt
Kashevarof Passage
N
SNOW PASSAGE

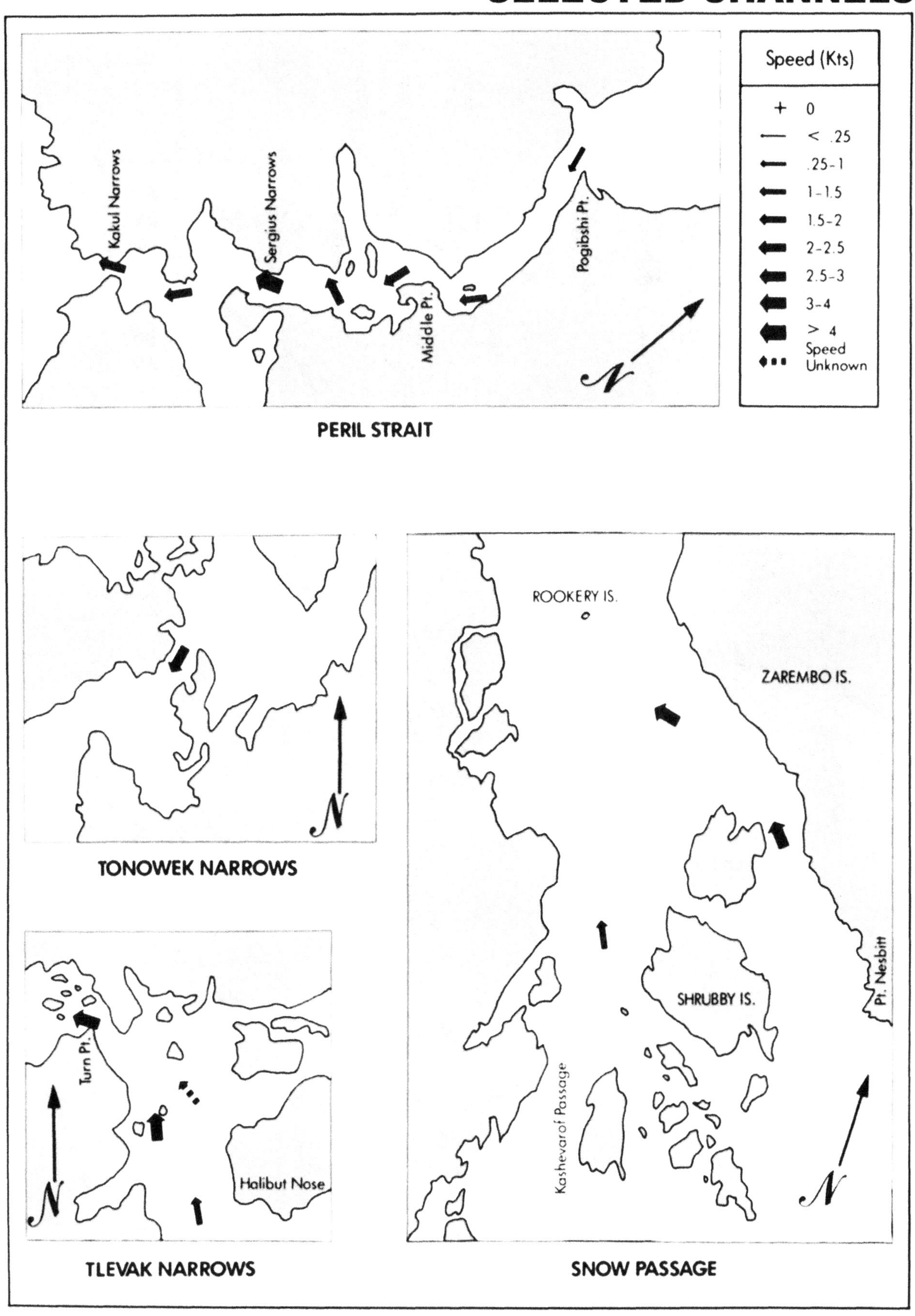
Speed (Kts)
+ 0
< .25
.25-1
1-1.5
1.5-2
2-2.5
2.5-3
3-4
> 4
Speed Unknown
Kakul Narrows
Sergius Narrows
Middle Pt.
Pogibshi Pt.
N
PERIL STRAIT
TONOWEK NARROWS
Turn Pt.
Halibut Nose
TLEVAK NARROWS
ROOKERY IS.
ZAREMBO IS.
SHRUBBY IS.
Pt. Nesbitt
Kashevarof Passage
SNOW PASSAGE

SELECTED CHANNELS

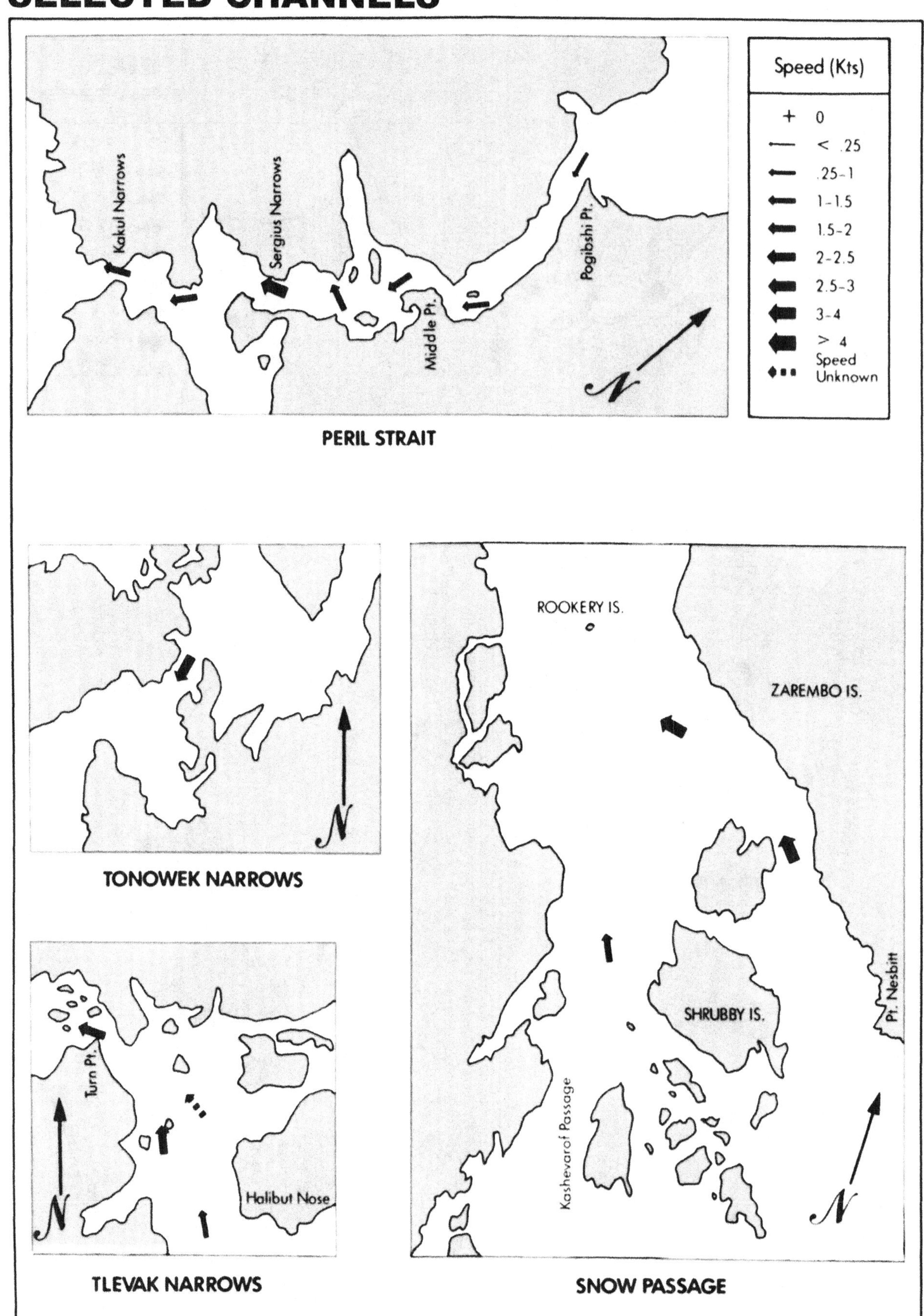

Speed (Kts)
0
< .25
.25–1
1–1.5
1.5–2
2–2.5
2.5–3
3–4
> 4
Speed Unknown
Kakul Narrows
Sergius Narrows
Middle Pt.
Pogibshi Pt.
N
PERIL STRAIT
N
TONOWEK NARROWS
Turn Pt.
Halibut Nose
N
TLEVAK NARROWS
ROOKERY IS.
ZAREMBO IS.
SHRUBBY IS.
Pt. Nesbitt
Kashevarof Passage
N
SNOW PASSAGE

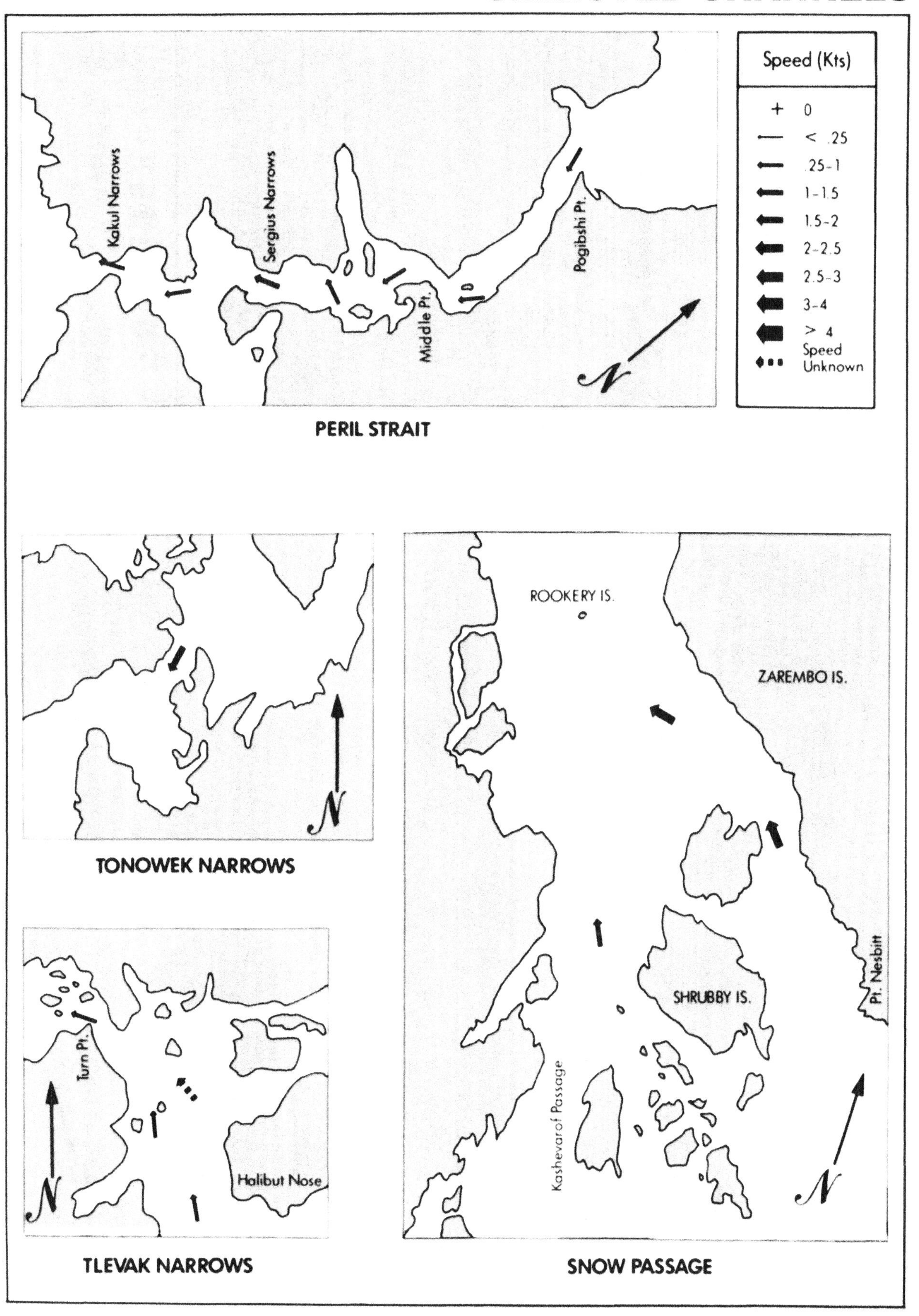
Speed (Kts)
0
< .25
.25-1
1-1.5
1.5-2
2-2.5
2.5-3
3-4
> 4
Speed Unknown
Kakul Narrows
Sergius Narrows
Middle Pt.
Pogibshi Pt.
N
ROOKERY IS.
ZAREMBO IS.
SHRUBBY IS.
Pt. Nesbitt
Kashevarof Passage
Turn Pt.
Halibut Nose

PERIL STRAIT

TONOWEK NARROWS

TLEVAK NARROWS

SNOW PASSAGE

SELECTED CHANNELS

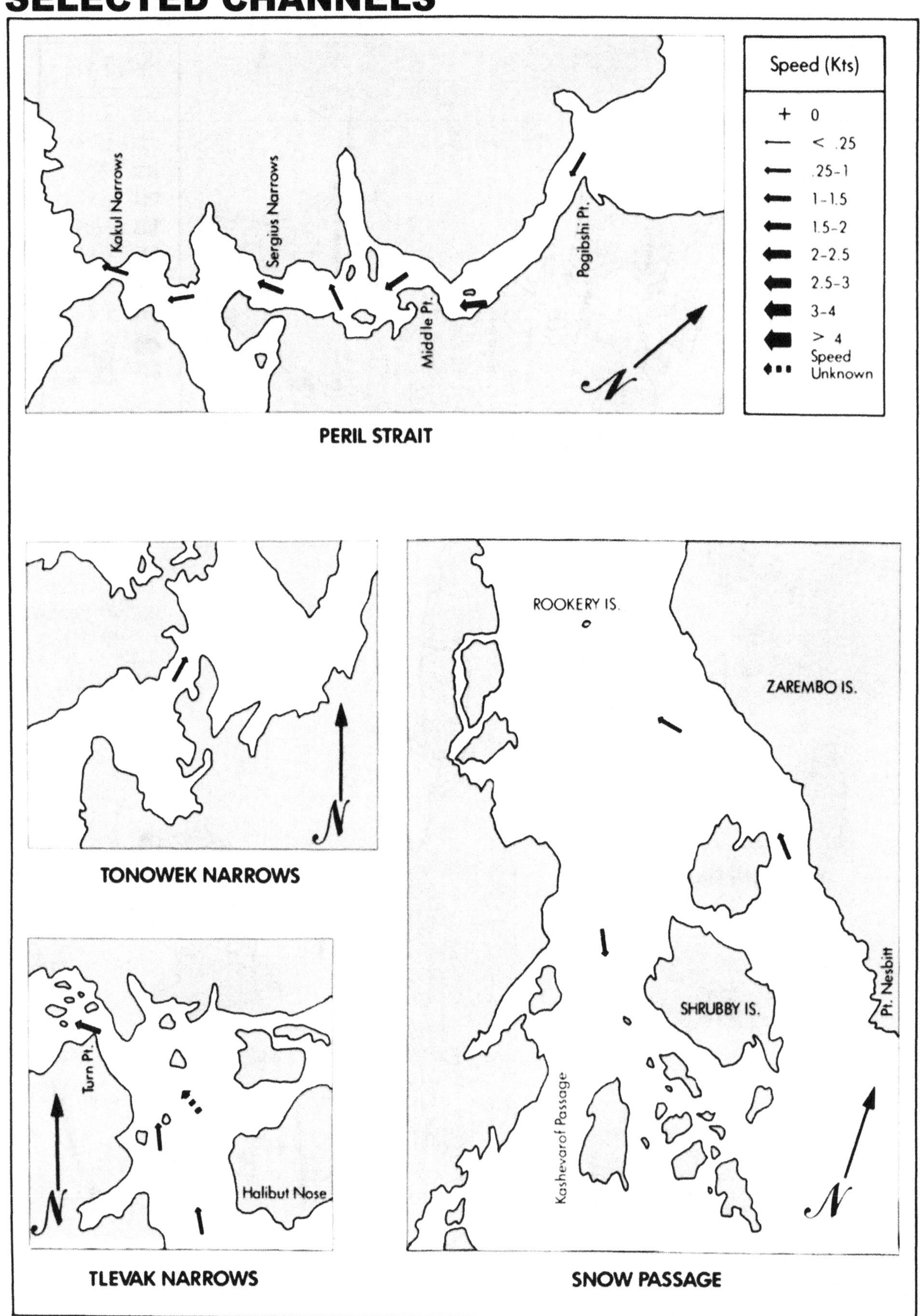

Speed (Kts)
0
< .25
.25–1
1–1.5
1.5–2
2–2.5
2.5–3
3–4
> 4
Speed Unknown
Kakul Narrows
Sergius Narrows
Middle Pt.
Pogibshi Pt.
PERIL STRAIT
TONOWEK NARROWS
Turn Pt.
Halibut Nose
TLEVAK NARROWS
ROOKERY IS.
ZAREMBO IS.
Pt. Nesbitt
SHRUBBY IS.
Kashevarof Passage
SNOW PASSAGE

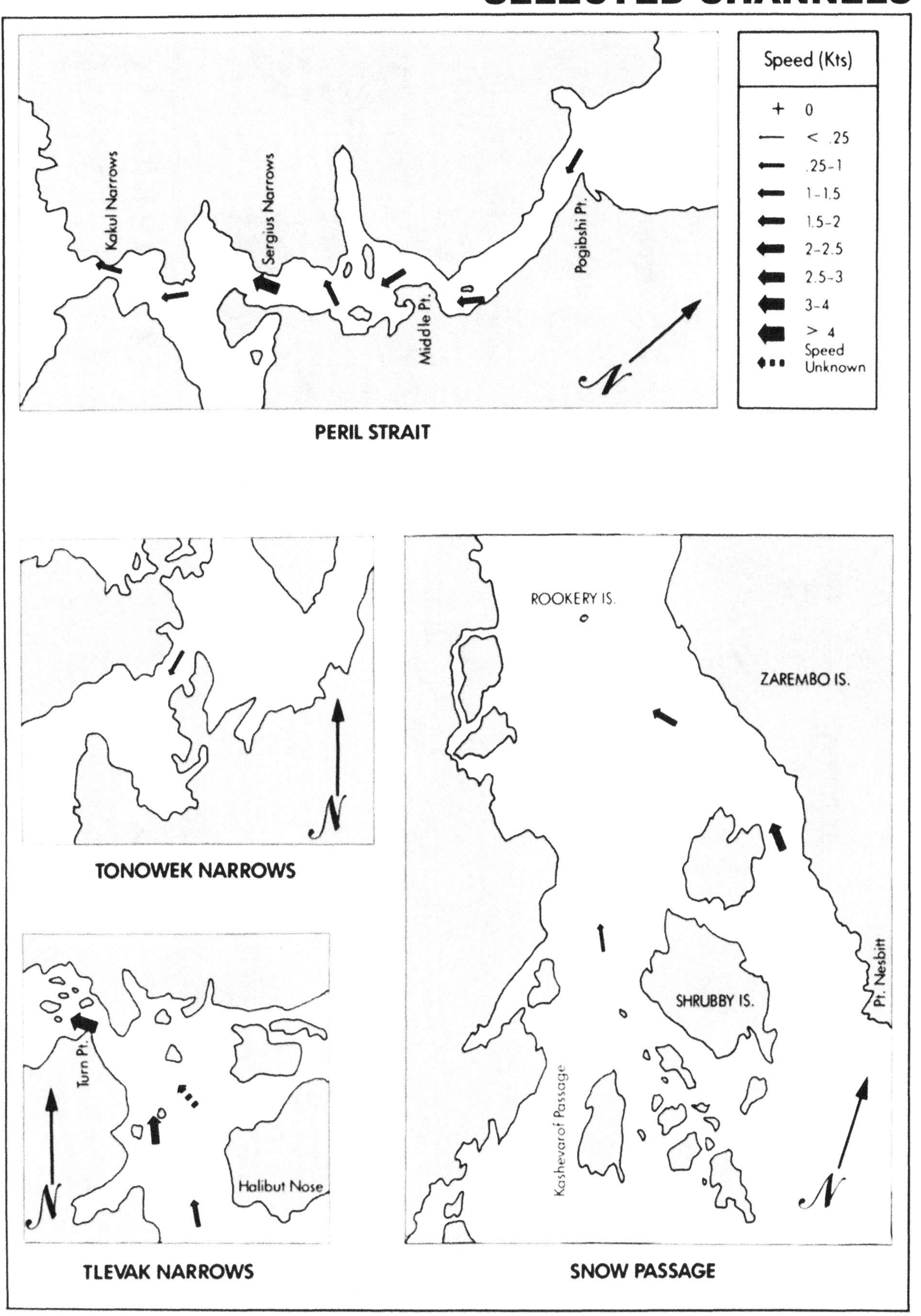
Speed (Kts)
+ 0
< .25
.25-1
1-1.5
1.5-2
2-2.5
2.5-3
3-4
> 4
Speed Unknown
Kakul Narrows
Sergius Narrows
Middle Pt.
Pogibshi Pt.
PERIL STRAIT
TONOWEK NARROWS
ROOKERY IS.
ZAREMBO IS.
SHRUBBY IS.
Pt. Nesbitt
Kashevarof Passage
Turn Pt.
Halibut Nose
TLEVAK NARROWS
SNOW PASSAGE

SELECTED CHANNELS

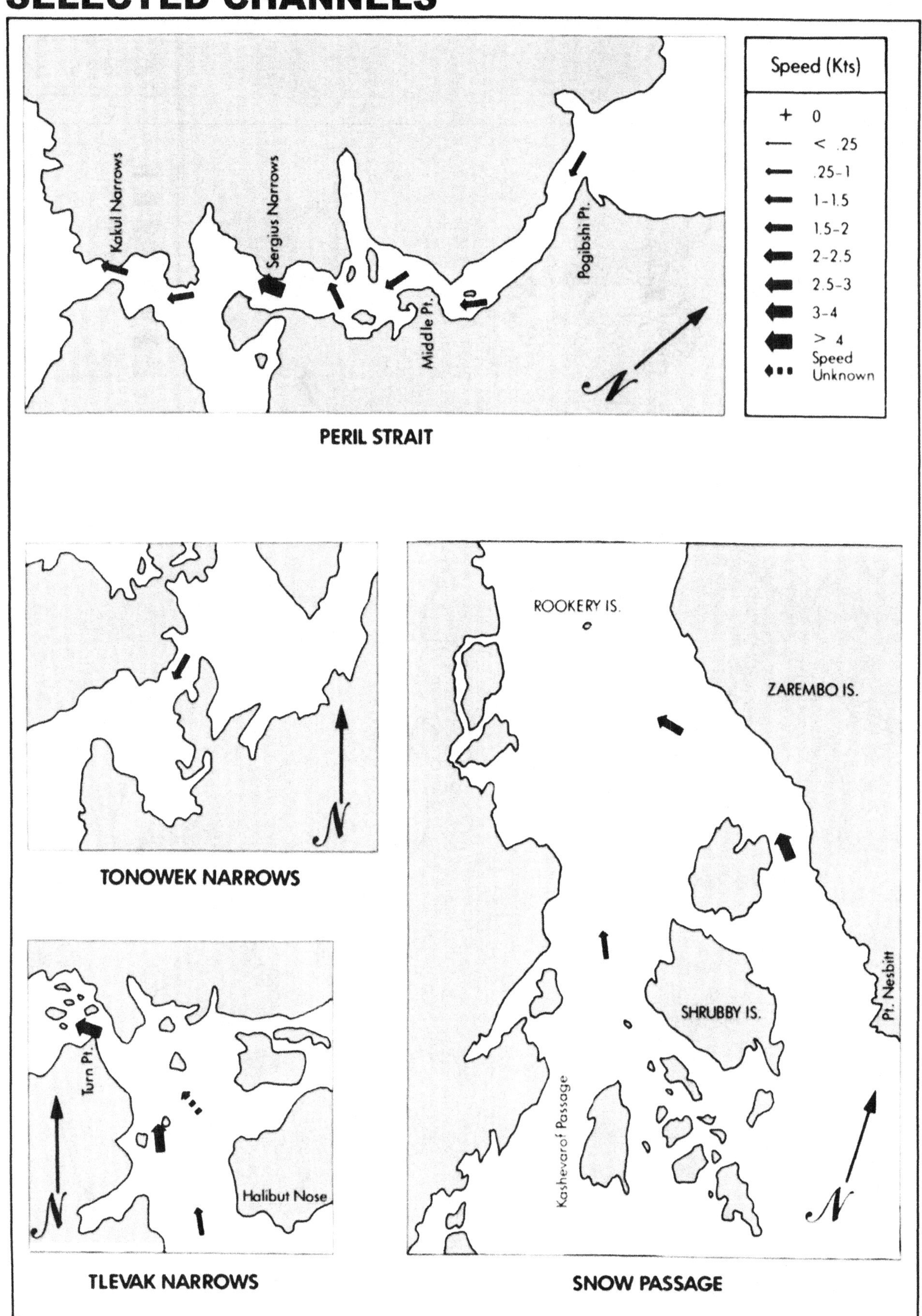

Speed (Kts)
0
< .25
.25–1
1–1.5
1.5–2
2–2.5
2.5–3
3–4
> 4
Speed Unknown
Kakul Narrows
Sergius Narrows
Middle Pt.
Pogibshi Pt.
N
PERIL STRAIT
TONOWEK NARROWS
Turn Pt.
Halibut Nose
TLEVAK NARROWS
ROOKERY IS.
ZAREMBO IS.
SHRUBBY IS.
Pt. Nesbitt
Kashevarof Passage
SNOW PASSAGE

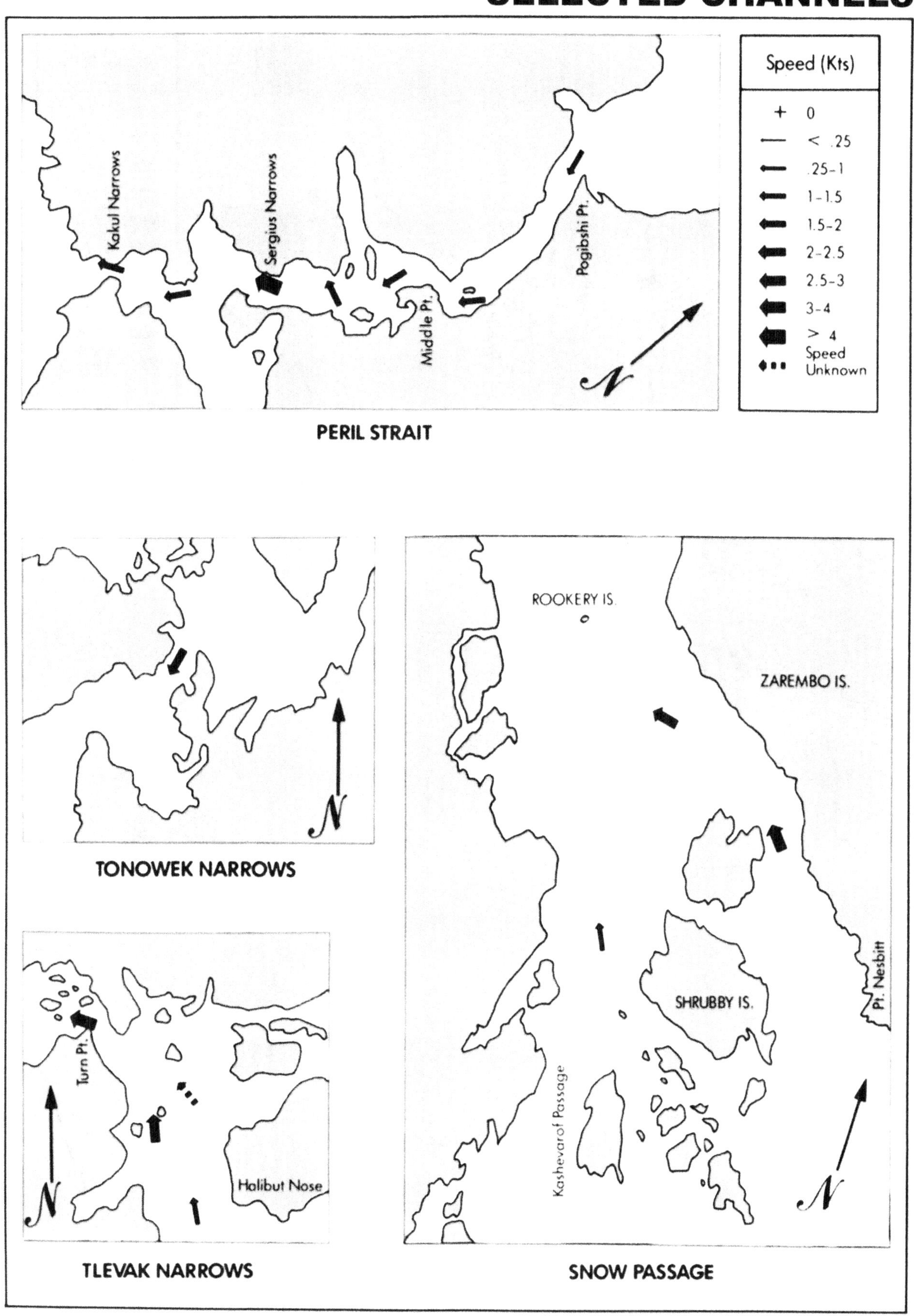
Speed (Kts)
+ 0
< .25
.25-1
1-1.5
1.5-2
2-2.5
2.5-3
3-4
> 4
Speed Unknown
Kakul Narrows
Sergius Narrows
Middle Pt.
Pogibshi Pt.
PERIL STRAIT
TONOWEK NARROWS
Turn Pt.
Halibut Nose
TLEVAK NARROWS
ROOKERY IS.
ZAREMBO IS.
SHRUBBY IS.
Pt. Nesbitt
Kashevarof Passage
SNOW PASSAGE

SELECTED CHANNELS

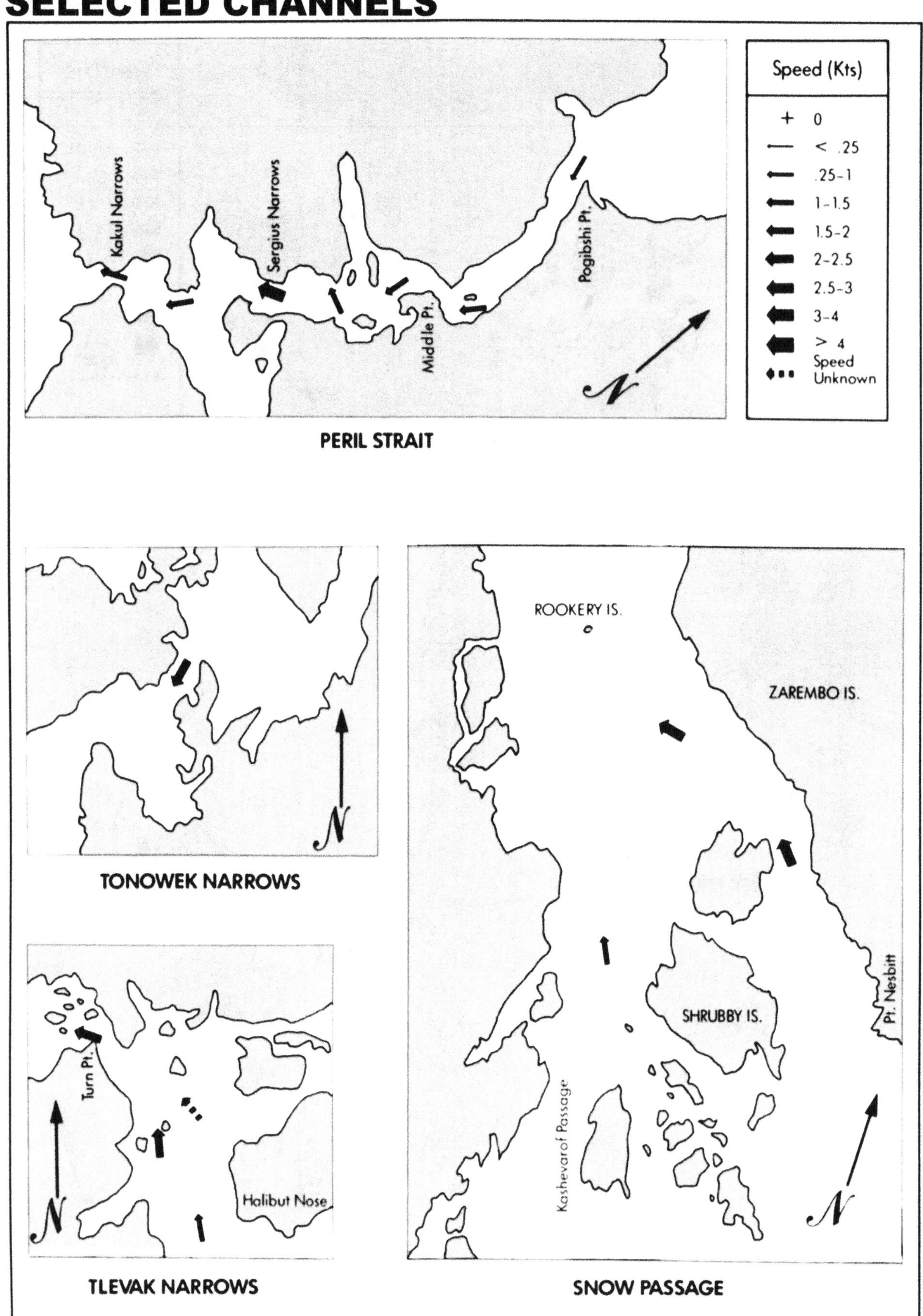

Speed (Kts)
+ 0
< .25
.25–1
1–1.5
1.5–2
2–2.5
2.5–3
3–4
> 4
Speed Unknown
Kakul Narrows
Sergius Narrows
Middle Pt.
Pogibshi Pt.
N
ROOKERY IS.
ZAREMBO IS.
SHRUBBY IS.
Pt. Nesbitt
Kashevarof Passage
Turn Pt.
Halibut Nose

PERIL STRAIT

TONOWEK NARROWS

TLEVAK NARROWS

SNOW PASSAGE

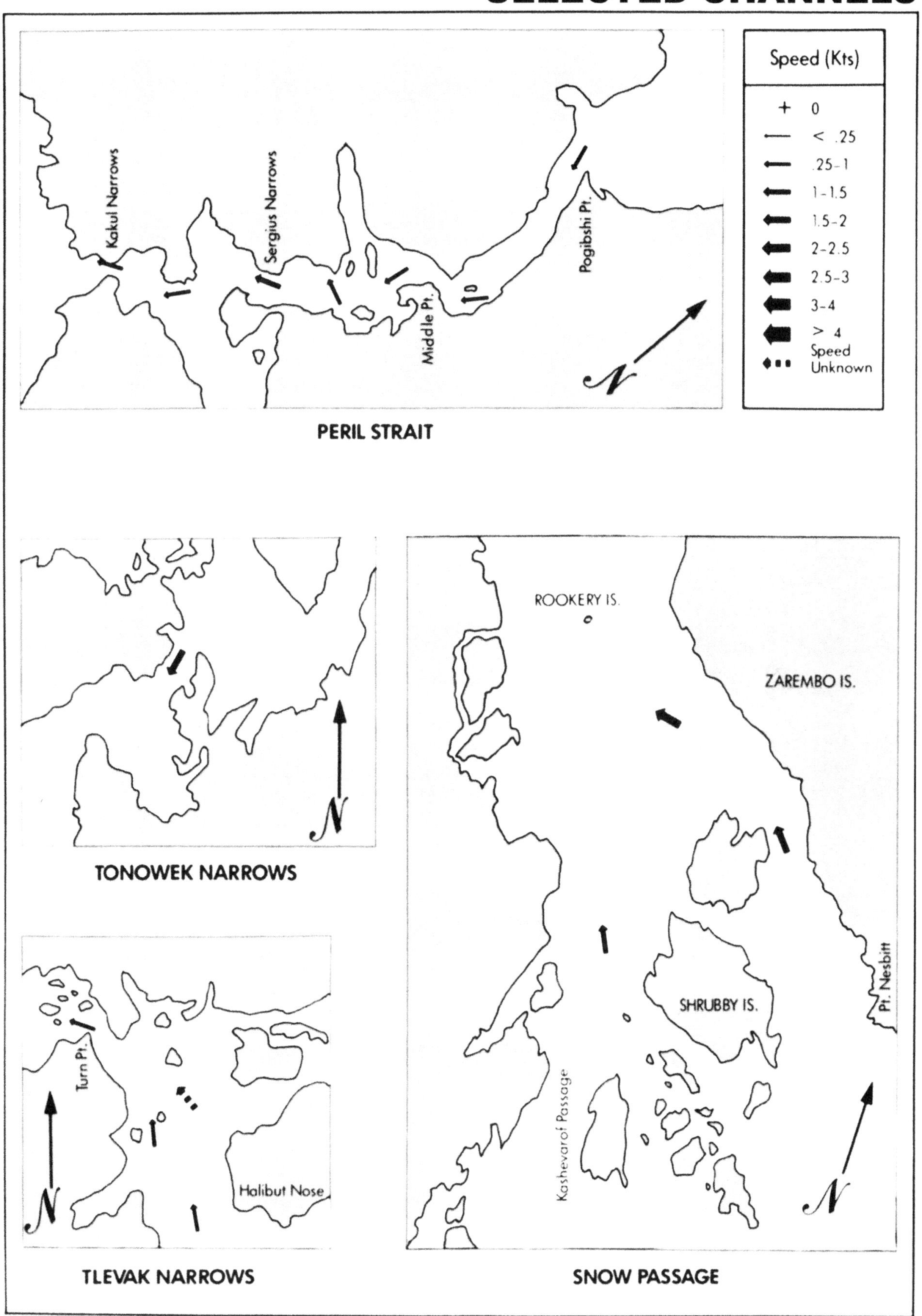
Speed (Kts)
+ 0
< .25
.25-1
1-1.5
1.5-2
2-2.5
2.5-3
3-4
> 4
Speed Unknown
Kakul Narrows
Sergius Narrows
Middle Pt.
Pogibshi Pt.
N
PERIL STRAIT
N
TONOWEK NARROWS
Turn Pt.
Halibut Nose
N
TLEVAK NARROWS
ROOKERY IS.
ZAREMBO IS.
SHRUBBY IS.
Pt. Nesbitt
Kashevarof Passage
N
SNOW PASSAGE

Related Titles Available from Starpath Publications...

Graphic current charts and flow patterns
TIDAL CURRENTS
of PUGET SOUND

Tidal Currents of Puget Sound
By NOAA, University of WA Dept. of Oceanography, and WA Sea Grant
ISBN 9780914025160

Paperback, 92 pages, 8.25" x 11".

Tide Prints show the flow patterns and how they evolve throughout the current cycle. There is one print for every 3 hours throughout the cycle. They are indexed to the tide height in Seattle.

Current Charts show the values of the currents at each reference station. There is one chart for every hour throughout the cycle. They are indexed to the tidal current at Admiralty Inlet.

Comparing Tide Prints and Current Charts shows the locations of eddies and bands of current, and how these bands and eddies move and interact as the current cycle evolves.

INLAND & COASTAL
NAVIGATION
SECOND EDITION
STARPATH
DAVID BURCH
For Power-driven and Sailing Vessels

Inland and Coastal Navigation, Second Edition
By David Burch
ISBN 9780914025405

Paperback, 228 pages, 8.25" x 11", many illustrations.

This book is an updated and expanded edition of a text that has been used in navigation courses for 30 years. It covers practical small-craft navigation (sail, power, or paddle), starting from the basics and ending with all that is needed to navigate safely and efficiently on inland and coastal waters in all weather conditions. It is for beginners, starting from scratch, or for more seasoned mariners who wish to expand their skills.

"An invaluable resource that belongs in the wheelhouse alongside Bowditch"
—Lee Chesneau, NWS Forecaster
MODERN
MARINE WEATHER
SECOND EDITION
From Time Honored Maritime Traditions to the Latest Technology
DAVID BURCH

Modern Marine Weather, Second Edition
By David Burch
ISBN 9780914025337

Paperback, 336 pages, 7.5" x 9.25", many illustrations.

A new, comprehensive text on how to take weather into account for the planning and navigation of voyages, local or global, using the latest technologies as well as the time-honored skills of maritime tradition, so that your time on the water remains as safe and efficient as possible.

It does not just tell you about it; it tells you how to do it.

CPSIA information can be obtained
at www.ICGtesting.com
Printed in the USA
LVOW04s1158130217
524067LV00018B/583/P

9 780914 025542